Descartes

To

Newton:

A Mentor's pre-High School Calculus Program

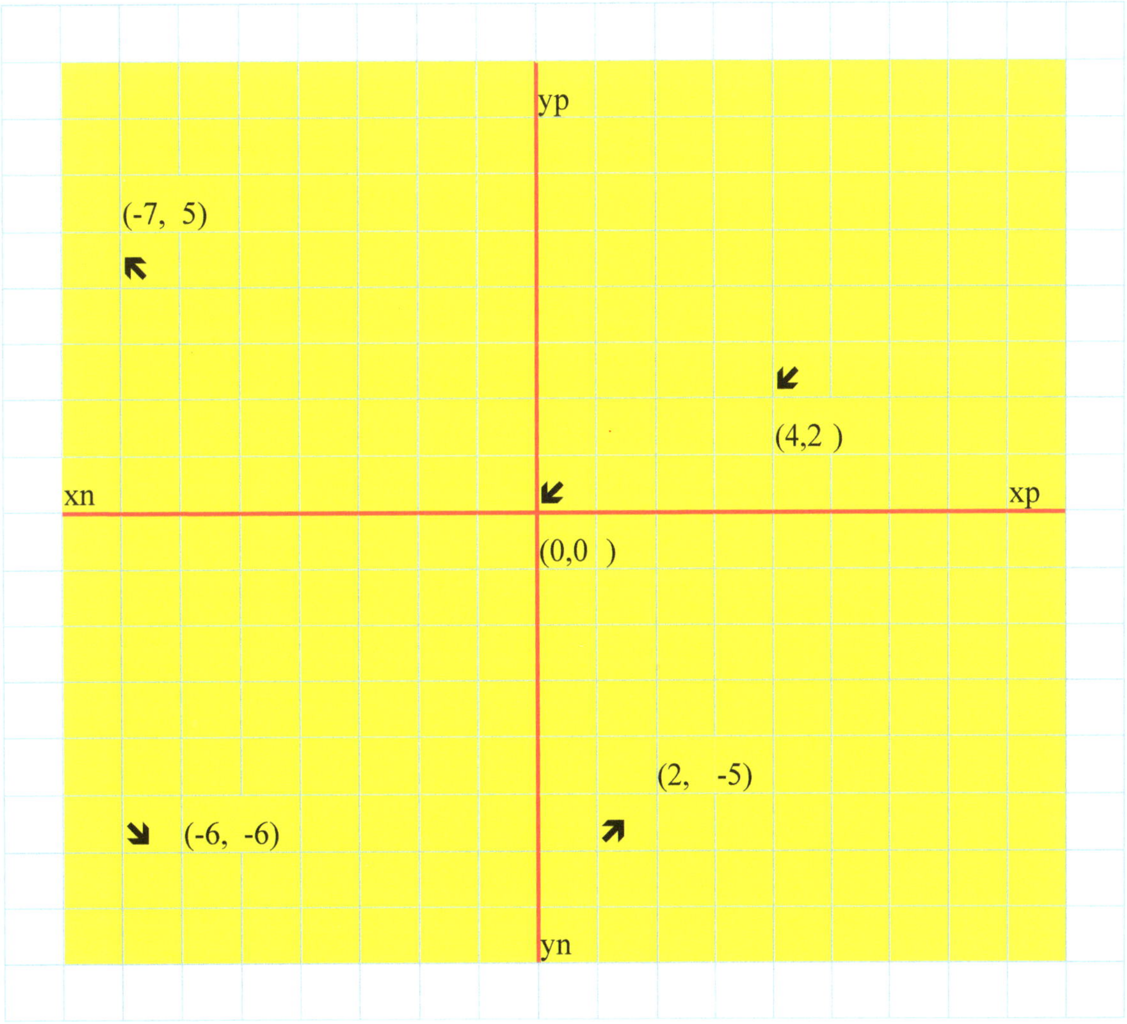

yp

(-7, 5)

(4,2)

xn   xp

(0,0)

(2, -5)

(-6, -6)

yn

Descartes To Newton:
A Mentor's pre-High School Calculus Program

The name Cuisenaire® and the color sequence of the rods
Are registered trademarks of ETA/Hand2Mind®

ISBN 10: 1508551022
ISBN 13: 9781508551027

Library of Congress control number: 2015902874

CreateSpace Independent Self-publishing Platform
North Charleston, SC

Printed in the United States of America

Dedication

To my wonderful wife, Sarah Emmons Warren, who has supported me through the ups and downs of writing a book along with all the other activities of a busy life.

Students Due Special Appreciation

A. Fortier, P. Hogg, E. Ney, M. Reid, E. Speck-Wayne, G. and U. Siegfried, C. Tiernan, all have continued my education in the thought processes of intelligent fifth and seventh grade children. Their questions have prompted me to look more closely at what I am teaching and how I am doing it. Their willingness to let me in on their view of mathematics has helped me see aspects of the subject more deeply.

A former middle school student of mine has to be mentioned at this point, C. Corey. The wonderful letter she wrote, reporting her progress at UMASS Amherst as a Computer Science and Mathematics major, encourages me to keep on mentoring young children in mathematics.

Acknowledgements

Thank you to those faithful readers, making the text so much richer for their suggestions. In particular, I want to express my thanks to Leonard Smith for his encouragement to look at the focus-directrix definition of the parabola. My thanks also to alarm-repairman 'David' who shared with me his problem-solving method (p29) and repaired my phone service in addition to the oil-tank system! Also, I would like to express my great appreciation of the numerous suggestions made by my former colleague, David A. Penner, who has read this text and made so many helpful comments. Another helpful critic is Tirzah Deering, whose knowledge of middle school children and their thoughts has helped me see problems through their eyes. Also, thanks to Louis Aikman for listening to my questions and making good suggestions on the topic of the history of Mathematics Education.

I alone am responsible for any remaining errors.

Introduction

Descartes to Newton is the result of a lifetime enjoying and teaching mathematics. I could hardly have thought of teaching middle school children calculus if I had not enjoyed learning mathematical topics, had not taught calculus at Phillips Academy, Andover, for thirty years, had not presented AP Calculus and AP Computer Science for the College Board for a dozen years, and finally had not spent ten years working with students in the Salisbury Elementary School or at the Boys and Girls Club of the Lower Merrimac Valley.

The young students of the Newburyport area demonstrated a genuine desire to understand mathematical ideas. My first arithmetic book, You Can Count On It: A Mentor's Arithmetic Patterns for Elementary Students has prepared fifth and sixth grade students to study analytic geometry, graphing linear and polynomial equations, calculating the effects of gravity on toy rockets, and finding areas under polynomials curves. The more I played math with young students, the more excited they and I became. An example is a sixth grade girl who can justifiably claim she has made intellectual progress and has gained due respect for her accomplishment. My delight in working with such students made me realize I had to write a report on my experience of aiming high. My teacher training experience for the College Board made me acutely aware of the mathematical needs of so many high school teachers, not to mention the needs of those in elementary school. Where teachers and their students can feel growth and empowerment, there tends to be an excitement and joy in the experience. An eagerness to push on in mathematics and other AP subjects seems to be the result of past accomplishments and an increasing awareness of the new opportunity.

The central purpose of this curriculum is to lead able young math students on an investigative path, discovering ideas they never thought of before, developing skills that empower them to go on to the next challenge. David Snowdon, in his Aging with Grace, uses the phase "idea density," reporting Sister Susan suggesting that idea density "depends on at least two important learned skills: vocabulary and reading comprehension … starting early in life." Snowdon adds: "most of the [brain's] growth comes during our earliest years." (p117) If reading to our children early in life prepares them for later verbal activity, why shouldn't we play with a greater density of math ideas (at least multiplication, division and order of operation) during the earliest years of school as well?

That scholarships, at both secondary school and college level, have been made available for the graduates of my program indicates others respect for the abilities of my former students, the first group at this writing in the third year of college.

It has been a privilege to work with the active young people I have had the good fortune to teach. They have taught me how much they can do, given the proper support and encouragement. I hope that both teachers and students will find a path to these exciting results that I report in my two books: You Can Count On It and Descartes to Newton.

Descartes to Newton
Table of Contents

Introduction

Act 1   Rene Descartes' Approach

Scene 1   The Cartesian Plane

Mentor: The French philosopher and discoverer, Rene Descartes, found a convenient way to convert geometric figures into numbers and vise versa. We have already seen how helpful our Cuisenaire Rods have been to help us understand arithmetic, and in Act I we will see how we can describe geometric figures—points, lines, polynomials, etc.— with numbers and equations. Our first exercise involves labeling the points in a Cartesian plane, picking two perpendicular lines and calling their intersection the <u>origin</u>, indicated by a pair of 0's as (0,0) or O(0,0). The bold horizontal line we call the x-axis; the bold vertical, the y-axis. We assign to every point in the plane a pair of numbers; the first indicates the horizontal distance from the y-axis, the second indicates the vertical distance from the x-axis. We think of the x-axis as pointing in the positive direction to the right, indicated by xp; the y-axis as pointing in the positive direction upward, by yp.

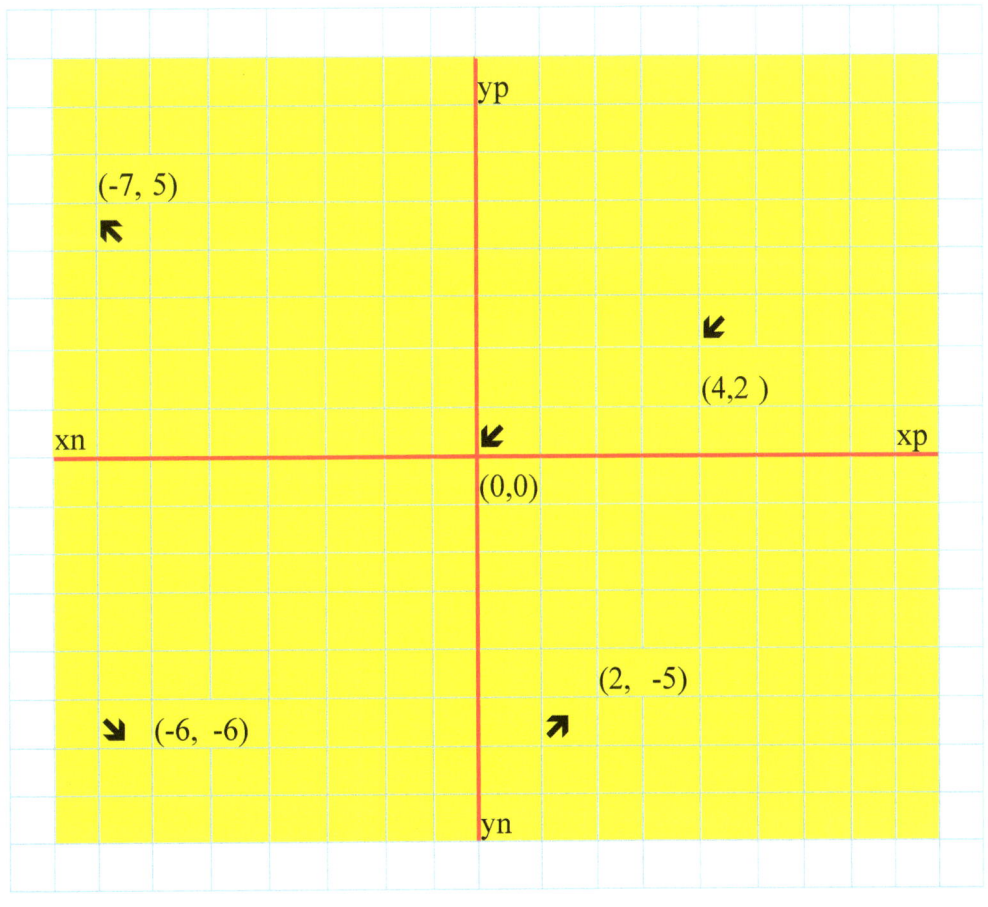

Figure 1

(Note: 1. All adjacent parallel lines are to be considered some common distance apart.
2. Pointers indicate the points of intersection of the two nearest perpendicular lines.)

Mentor: What are the coordinates of the indicated points?

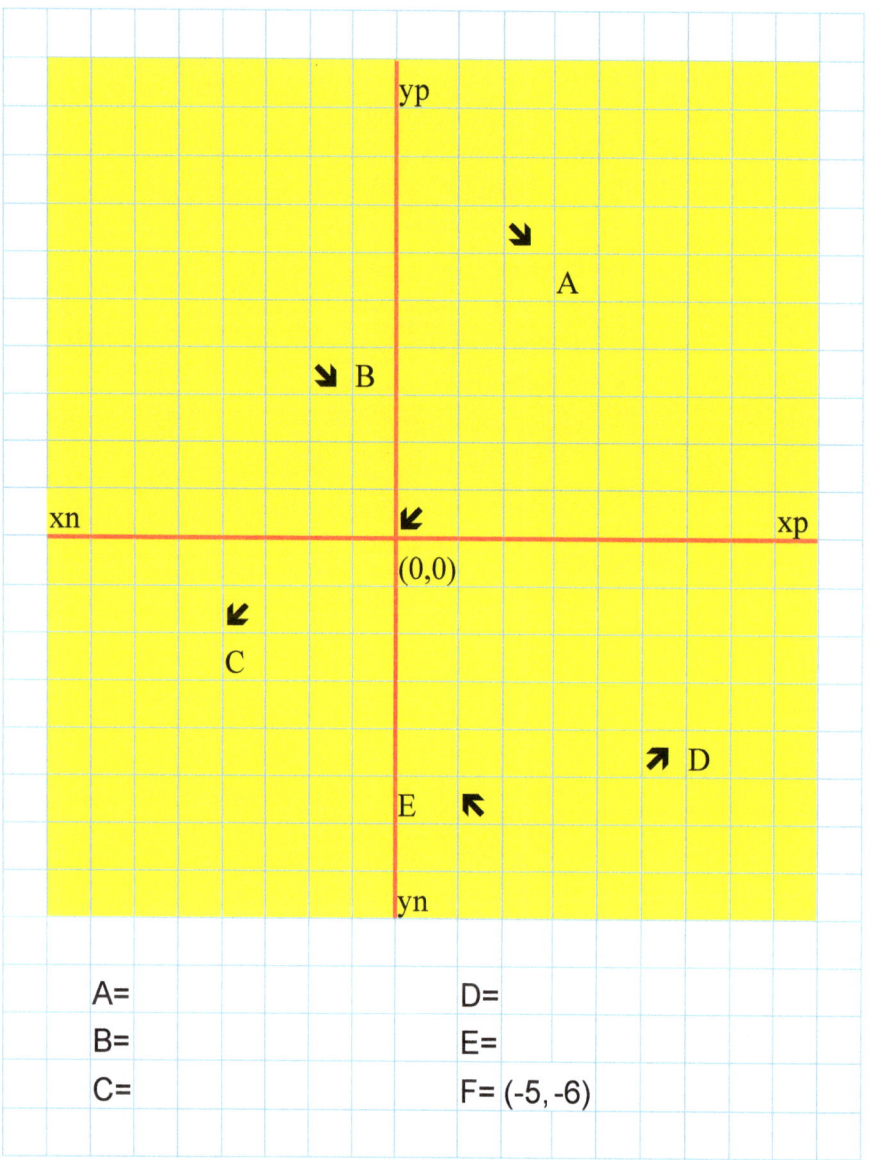

Figure 2

Beatrix: A = (3, 6).   **Emma:** B = (-1, 3).   **Macey:** C = (-4, -2).
Maddy: D = (6, -4).   **Ursula:** E = (1, –5).

Mentor: We see that the coordinate axes divide the plane into four regions, called quadrants. The quadrant above the x-axis and to the right of the y-axis is the first quadrant, indicated by the Roman numeral I. The quadrant on the other side of the y-axis we call the second quadrant, II, the third quadrant, III, is below the second, and the fourth, IV, completes the plane. Where is the point F?

Georgia: (-5, -6) is a point in the third quadrant 5 units to the left of the y-axis and 6 units down from the x-axis. So, in quadrant III!

Mentor: Where are all the points with x-coordinate 6, or x = 6?

Beatrix: $x = 6$ for all the points on the vertical line through point D in Figure 2.

Mentor: Where are all the points with y-coordinate -2, or $y = -2$?

Ursula: $y = -2$ describes the horizontal line two units down from the x-axis.

Mentor: Excellent. Now we can look for some more geometric figures described by pairs of numbers.

Between the Scenes

1) Describe each of the lines passing through each of the named points in Figure 2 and give the equation of each line. Example: The two lines passing through the origin, O (0, 0), are the coordinate axes: $x = 0$ is the equation for the y-axis and $y = 0$ is the equation for the x-axis.

2) **Caroline:** What is this 'x' in $x = 1$?

Mentor: What's in a box labeled 'SOAP' ?

Caroline: I suppose soap.

Mentor: Suppose the box contained sand or toy cars or tennis balls.

Caroline: Then the label on the box would be misleading.

Mentor: What would you say about my statement "The box labeled 'SOAP' contains soap," if the box actually contained dolls or toy trains?

Caroline: Your statement would be false! What does this box have to do with my question about $x = 1$?

Mentor: If you think of 'x' as a label on a box, what do you expect to find in the box?

Caroline: I expect to find 1 in the box.

Mentor: What if I told you there was a 2 in the box?

Caroline: Then the statement would be false!

Mentor: Is the statement "Her name is Susi" true or false?

Caroline: If the 'her name' refers to me, the statement is false, because my name is Caroline. If it refers to my mother it would be true because Susi is her name.

3) **Mentor:** What is a simplified way of writing $-(-3)$? Of $-(-x)$? What makes it simpler?

Act 1   Scene 2   Equal Coordinates

Mentor: Now that you know where to find a point from its coordinates, let us consider where a collection of points will be if they satisfy some given condition. First, let's consider all the points with their x- and y-coordinates equal.
Beatrix: I know one such point: the origin.
Macey: We have another on our first graph: (-6, -6) in quadrant III.
Emma: We can name a bunch of these points: (1, 1), (2, 2), (3, 3). They run up a diagonal from the origin to the upper right-hand corner of our graph.
Mentor: You all have the idea. Looking at their graphed points, what pattern do you see?

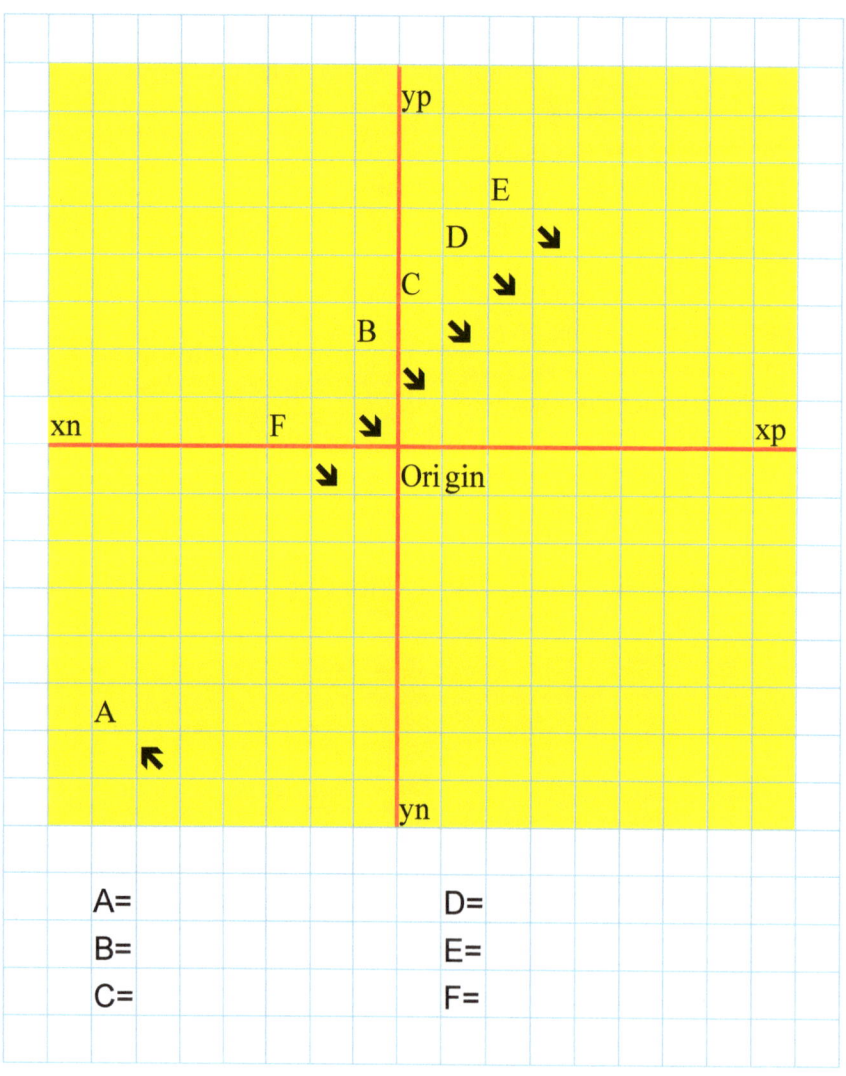

Figure 3

Mentor: What geometric figure do you see all the arrows pointing to?

4

Maddy: I see a straight line.

Mentor: Yes. What equation expresses the relation between the x- and y-coordinates?

Ursula: y-coordinate = x-coordinate.

Mentor: Would the removal of the repeated word "coordinate" simplify the expression?

Ursula: Yes. The equation would then read: y = x.

Mentor: Suppose the x-coordinate on this figure was a fraction? Say ½?

Emma: Then the y-coordinate would have to be ½.

Mentor: So all the points with equal coordinates would lie on this line including those with fractional values. Let's draw the line.

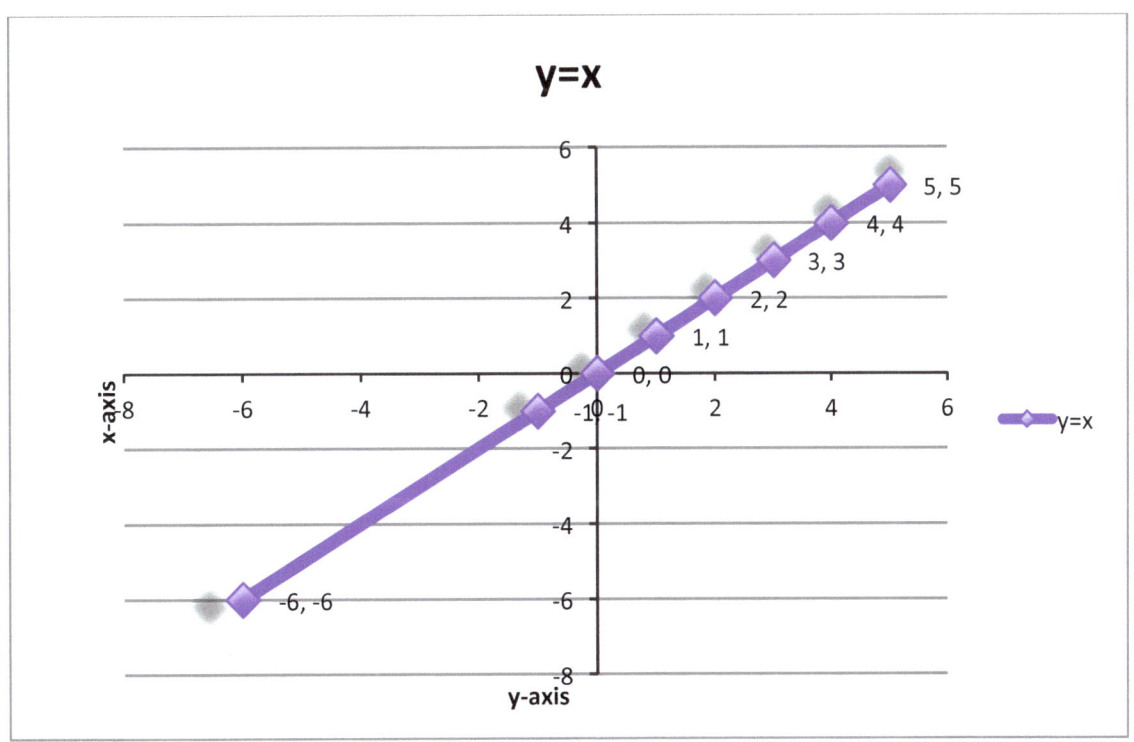

Figure 4

Mentor: That is excellent. In the next scenes we will consider what happens to the graphs of the line when we add some number to the x side (y = x + 2), or subtract some number from the x side (y = x – 6), or multiply or divide the x side of the equation by some number (y = 5x or y = x/8, respectively).

Between the Scenes

Try graphing each of the suggested lines in this last Mentor comment.

Act 1   Scene 3   Adding and Subtracting

Mentor: Now let's consider the graph of all points where the y-coordinate is two more than the x-coordinate. Who could name some of those points?

Ursula: (0, 2), (1, 3), (2, 4), (3, 5), (-6, -4).

Mentor: Where is each of these points relative to those with the same coordinates on our first line?

Ursula: Moved two units up on the y-axis.

Mentor: Good. Show us the graph. Note that the software provides the coordinates of all the points on our original list but does not provide parentheses around them. We should continue to use the parentheses. So we will write (5,7) instead of just 5,7.

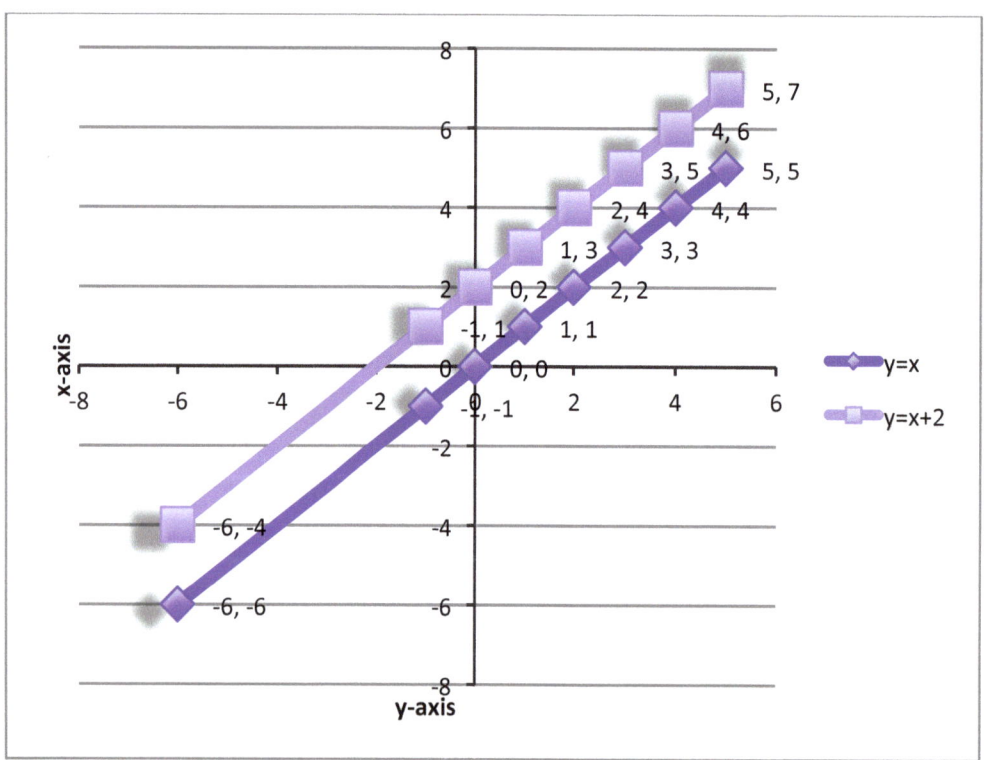

Figure 5

Mentor: Seeing that change on Figure 5, what will the graph of y = x − 3 look like?

Maddy: We will move the original line down three instead of up two as in Figure 5. Now the graph looks like this. In Figure 6 in lightest purple is the graph we wanted.

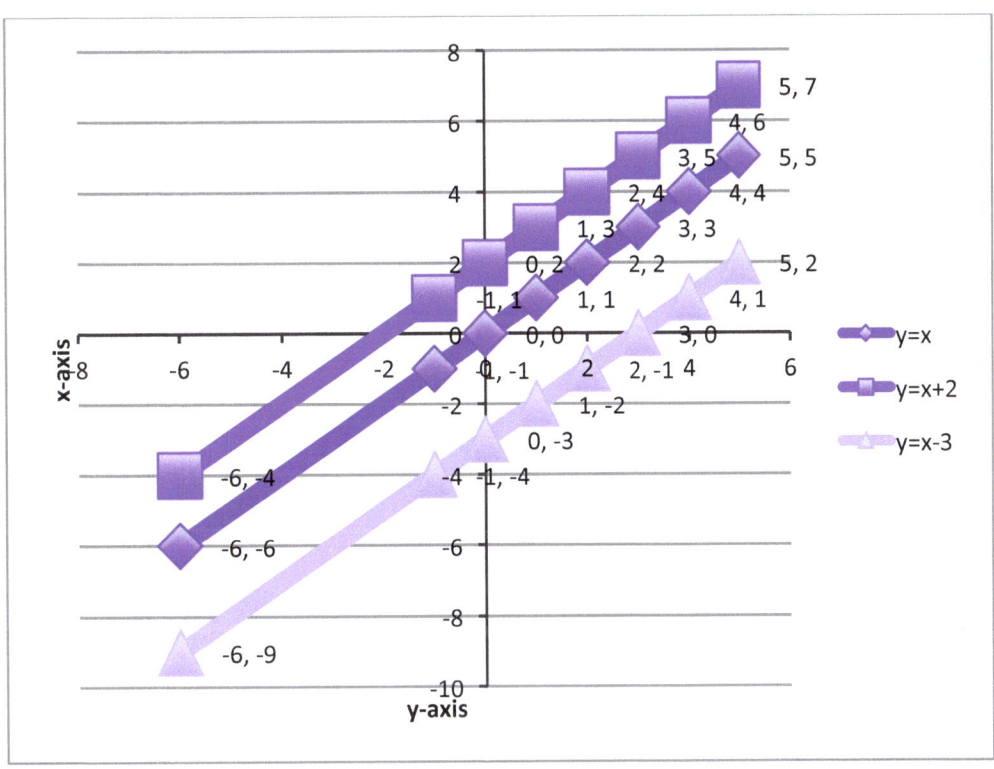

Figure 6

Mentor: Who can state for us what change we make to the graph of the line y = x by adding or subtracting a constant value to the x side of the equation?
Maddy: We have seen that if we add a positive number to x to get the new y value, we will raise the graph of y = x by that positive number. Similarly for subtracting a positive number, we lower the given line by that number.
Mentor: Will these lines ever intersect?
Maddy: No! They are parallel to each other.
Mentor: Now we move on to the next aspect of lines and their equations.

Between the Scenes
Start with the point A(5, 5) on line y = x.
  (a) What will the y-coordinate of the point B(5, y) equal on the line of y = x + 2?
  (b) What will the y-coordinate of the point C(5, y) equal on the line of y = x – 3?
  (c) What will the y-coordinate of the point D(5, y) equal on the line of y = x + 7?
  (d) What will the y-coordinate of the point D(5, y) equal on the line of y = x – 7?

Act 1   Scene 4   An Opposite Interlude

Mentor: While we have the equations and their graphs in front of us, let us look at the process of finding x-coordinates given y-coordinates. On the line of y = x + 2, what is the x-coordinate where y = 0?

Maddy: I see that where y = 0 the line crosses the x-axis at x = -2.

Mentor: So how do we solve the equation 0 = x + 2 without looking at the graph?

Maddy: Only by adding negative two to two can I get zero.

Mentor: Excellent. What would you have to do to both sides of 0 = x + 2 to leave the x all by itself on the right?

Piper: I would subtract 2 from the right: x + 2 – 2 = x.

Mentor: Good, but do you still have an equation? What would it be?

Piper: If I took 2 away from the right side without doing anything to the left, I would have 0 = x. But x does not equal zero, rather -2. So I guess I should subtract 2 from the left side when I subtract it from the right. Now I have:

$$0 - 2 = x + 2 - 2,$$
$$\text{or}$$
$$-2 = x,$$
$$\text{or}$$
$$x = -2$$

Mentor: Nicely worked out! Can somebody help us with the equation: y = x – 3, where y = 0? Or 0 = x – 3?

Georgia: That's easy! x has to be 3 if we subtract 3 and get zero!

Mentor: Nice! Tell us another approach, keeping the equation in focus.

Georgia: Oh, I see. I will add 3 to both sides of the equation:

$$0 + 3 = x - 3 + 3,$$
$$\text{or}$$
$$3 = x,$$
$$\text{or}$$
$$x = 3$$

Mentor: Yet another question to extend the observation. If y equals another number other than zero, say 8, how would the solution process go?
$$8 = x + 2?$$

Noel: I would subtract 2 from the left and right sides, giving us:
$$8 - 2 = x + 2 - 2,$$
$$\text{or}$$
$$6 = x,$$
$$\text{or}$$
$$x = 6$$

Mentor: And if we have $8 = x - 3$?

Noel: Then I add 3 to both sides!

$$8 + 3 = x - 3 + 3,$$
$$\text{or}$$
$$11 = x,$$
$$\text{or}$$
$$x = 11$$

Mentor: Who would like to state what we have just seen? Can we state a rule covering both cases?

Noel: I've got it! If a number equals x added to some second number, then we subtract the added number from both sides of the equation, and if a number equals x minus some second number, then we add the subtracted number. Whenever we want to remove the added or subtracted number from the side with x, we subtract or add it, so it will cancel out!

Mentor: That is our story of opposites. To undo an addition, subtract; to undo a subtraction, add.

Between the Scenes

Reviewing your work with the Cuisenaire Rods and other means of calculation, find the value of $65 - 27$. The 'traditional' way I was taught had me 'borrowing 10 from the 60 to make subtraction of 7 from 5 possible. That was working from right to left. Using the Rods didn't suggest that approach to me. How did you solve the problem? Explain each of the processes you used to arrive at your answer. Use your methods to calculate the following differences.

a) $71 - 35$ b) $43 - 26$ c) $92 - 78$ c) $54 - 16$

Act 1   Scene 5   Multiplying and Dividing by a Constant

Mentor: Now let us consider what happens to the line y = x, if we multiply the x-coordinates by 2, giving us the equation y = 2x. What are the coordinates for the new points?
Ursula: (-6, -12), (-1, -2), (0,0), (1, 2), (2, 4), (3, 6), (4, 8), and (5, 10). So Figure 7 shows the new line.

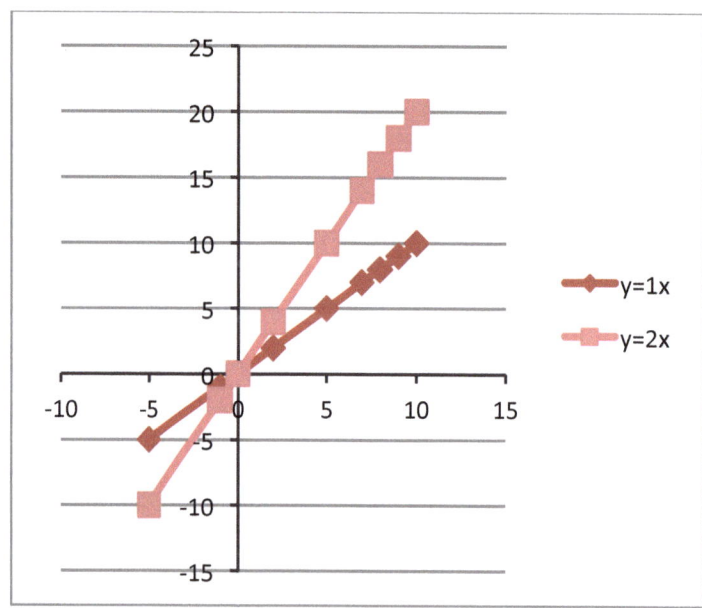

Figure 7

Mentor: Now tell me what has changed and what has not.
Emma: The origin has stayed put while the rest of the line seems to have rotated around the origin, with the positive side spinning up and the negative side spinning down.
Mentor: That description is a good start. Now let's consider some individual points. This graph does not label each data point, but we can read from the coordinate axes what the coordinates of the points are. Going out to the second point to the right of the origin, we find the red diamond point at (5,5). Which pink point corresponds with this point?
Emma: The point (5,10).
Mentor: Yes. That is correct. If we had really rotated the line around the origin, would the point (5,5) have ended up at (5,10)?
Maddy: I know! When we multiplied by 2 we doubled the height above the x-axis. So we are actually stretching the line out by doubling the height of the original point.
Mentor: Excellent. Measuring by eye, we can see that the point (5,10) is twice as high as the point (5,5) but has not moved off the vertical line x = 5. What do you think will happen if we were to multiply x by -3 instead of by 2?
Trix: I see it! The heights would be 3 times as great and in the opposite directions!
Mentor: Let's draw it.

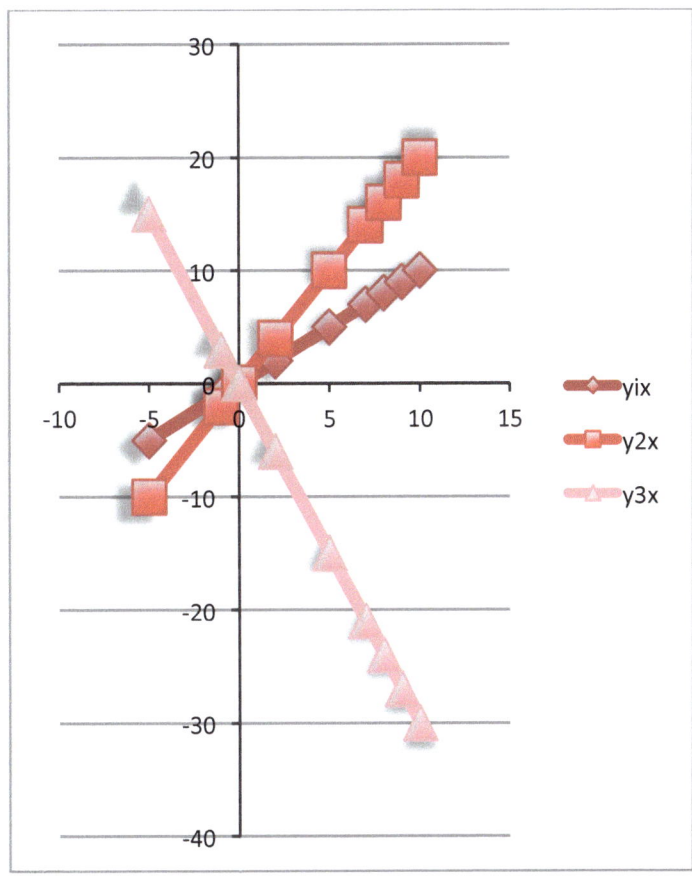

Figure 8

Trix: It is just as I said. The height of the red square point (10, 20) is twice as far from the x-axis as the diamond point (10,10). Also the lightest pink point (10, -30) is three times as far from the x-axis but 30 units down below the x-axis.

Mentor: That is good observation for the positive side of the x-axis, but what happens to the points on the left side of the y-axis?

Ursula: The minus sign flips the points that are above the x-axis down below it and the points below the x-axis up above it.

Mentor: In our last variation we considered dividing x by 2, or multiplying by ½. What happens to y = x, if we change the equation to $y = \frac{1}{2}x$?

Ursula: This line still goes through the origin, but the line goes up more gradually. To go up one unit on the y-axis, we have to go over 2 units on the x-axis. The negative side of the line is in the third quadrant, between the negative y-axis and y = x.

Mentor: I sense that we are in control of these ideas. Let's look at a new definition we will find very useful whenever we use the Cartesian plane.

Act 1   Scene 6   Another Story of Opposites

Mentor: If we consider the equation of a line, say y = 2x, and want to find the value of x if y = 1, how would you go about doing it? Note that if y = 0, x will be equal to 0 also.

Trix: I think it would be like the addition-subtraction situation.

Mentor: OK! So what would you do to solve the equation: 1 = 2x?

Trix: I recognize that two halves make one whole, so x must equal $\frac{1}{2}$. However, I also see that if I divide both sides by 2, then I get the equation $\frac{1}{2} = x$.

Mentor: So have you worked out a rule for us to follow?

Trix: Yes! If y is 1 and equals a constant times x, then we solve for x by dividing by the <u>coefficient</u> of x.

Mentor: So, how do you solve 1 = 7x?

Trix: I divide both sides of the equation by the coefficient, 7, and I get: $\frac{1}{7} = x$.

Mentor: So far, so good. Now suppose we have y equals 1, and x is divided by some number. Try 1 = x/5.

Piper: This time we respond to division with multiplication, its opposite. So multiply by 5, giving us $5 \times 1 = x$, or 5 = x.

Mentor: Now suppose we get something like 3 = 5x?

Piper: That is easy. All I have to do is divide by 5 on both sides. Because 5 divided by 5 will give us one, we will have simply x on the right. So we are left with:

$$\frac{3}{5} = x,$$
$$\text{or}$$
$$x = \frac{3}{5}$$

Mentor: Would anybody like to tell me a general rule for solving the equation with a constant on one side and x times a non-zero number on the other?

Piper: Sure. We will divide both sides by any number multiplied times x.

Mentor: What will we do if a number divides x?

Piper: If a number divides x, then we will multiply both sides by that number.

Mentor: Solve the following equation: 6 = x/11.

Georgia: I multiply by 11, to get x = 66.

Mentor: What is the general rule for solving such equations?

Georgia: Use the opposite operation to remove the constant from the x side of the equation.

Mentor: Once again we find our rule that has us perform the "opposite" operation to solve the equation. For the equation 3 = 8x, we divide by 8 to get x = 3/8. For the equation 3 = x/8, we multiply by 8 to get x = 24.

As our expressions in various equations become more complicated we have to have the methods of simplifying them.

Act 1   Scene 7   Rate of Climb

Mentor: In this scene we will introduce the so-called <u>slope</u> of a line.

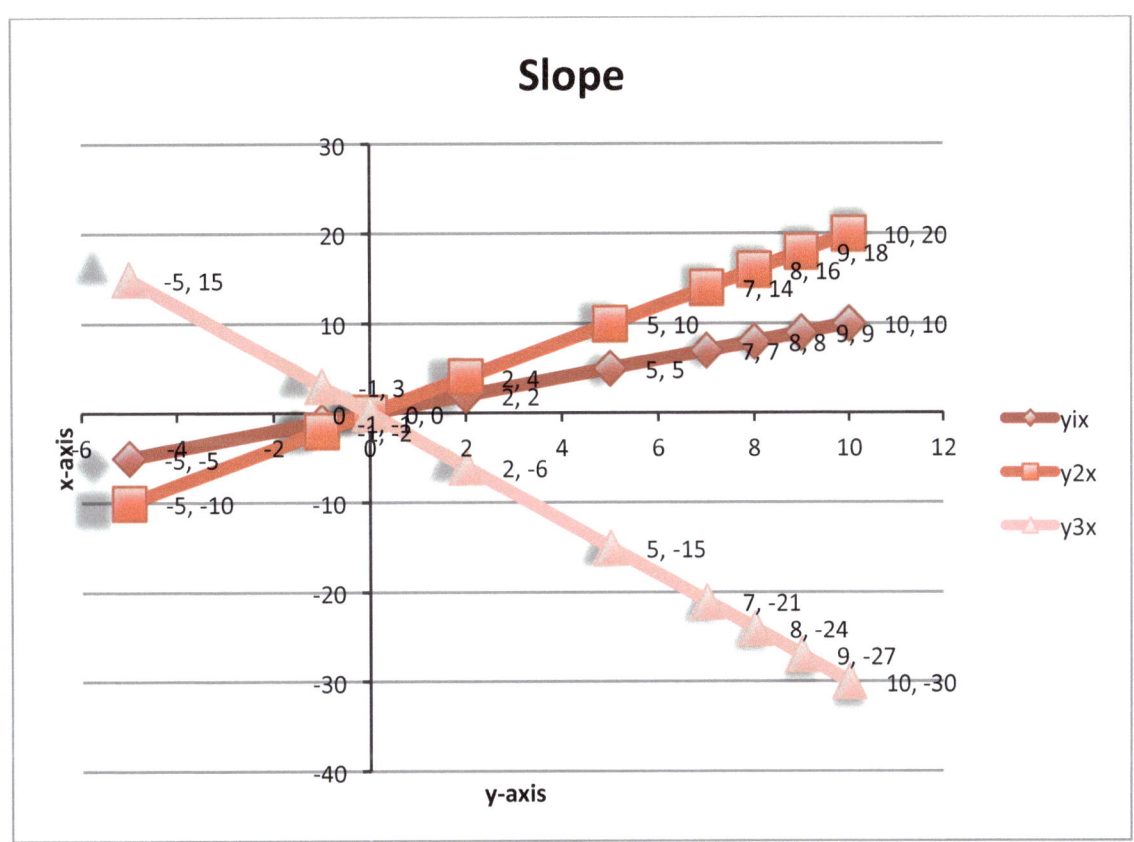

Figure 9

Mentor: If we start at the origin, O (0,0), and find the point on the x-axis vertically below the point (5,5), what will the coordinates of this point be?

Emma: It will be (5,0), because the y-coordinate of any point on the x-axis is zero and all of the points on the vertical line passing through (5,5) and (5,10) and (5, -15) have x-coordinates of 5. So x = 5.

Mentor: So let's draw the triangle, in Figure 10, with the three corners, called <u>vertices</u>, at O (0,0), A(5,0) and B(5,5). And let us form the ratio of the length of the vertical side, AB, divided by the length of the horizontal side, OA. What do we get?

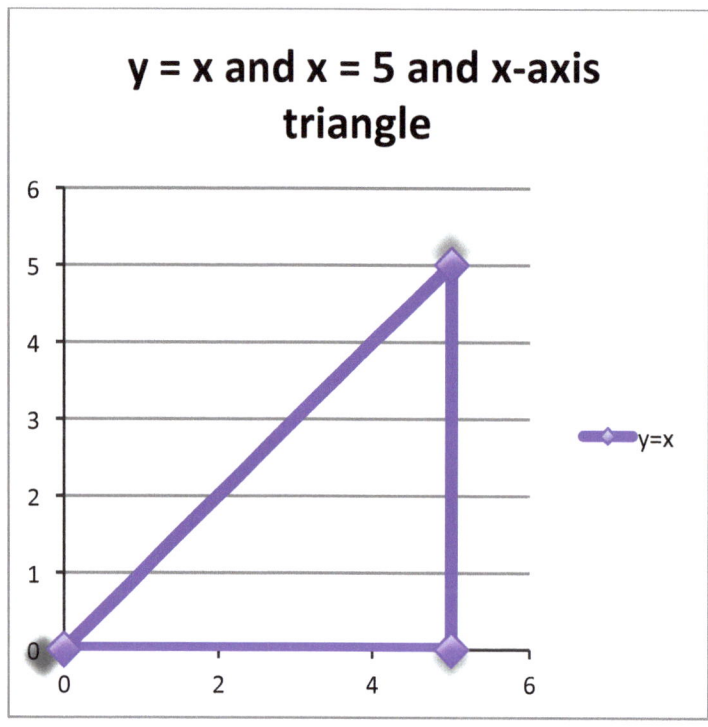

Figure 10

Maddy: The ratio is 1.

Mentor: How did you arrive at that number?

Maddy: Ratio = $\frac{5-0}{5-0}$.

Mentor: We can easily see that the height, we call it the 'rise,' over the horizontal distance, we call the 'run,' equals 1. We call this number the <u>slope</u> of the line y = x. What would we get for the slope of y = 2x?

Emma: The new ratio is formed by a new triangle with the vertex at the top at (5,10). So our ratio for the slope would be $\frac{10-0}{5-0} = 2$.

Mentor: Now suppose our triangle had one side on the y = -3x line. What would we get for a slope?

Trix: This is easy! The vertical line, x = 5, intersects the line y = -3x at the point (5, -15), so our ratio is $\frac{-15-0}{5-0} = -3$.

Mentor: Does anybody see a pattern developing here?

Ursula: Oh, yes! The number we multiplied times the x-coordinate is the slope!

Mentor: If we were traveling from the origin in the direction of the positive x-axis along a line with a positive slope, would the y-coordinates be going up or down?

Ursula: A positive slope means we would be rising up from the origin if we ran out the positive x-axis.

Mentor: Then does it make sense that we sometimes describe the slope as the "rise over the run?" We might see this as the path of an airplane climbing out of an airport; the slope indicated how high the plane would be for a given horizontal distance. Stay tuned!

Act 1   Scene 8   Slope Calculations

Mentor: In this scene we will show where four lines intersect the axes and find the slope of each line. Looking at Figure 11, calculate the slope of the line through (0,4) and (7,0).

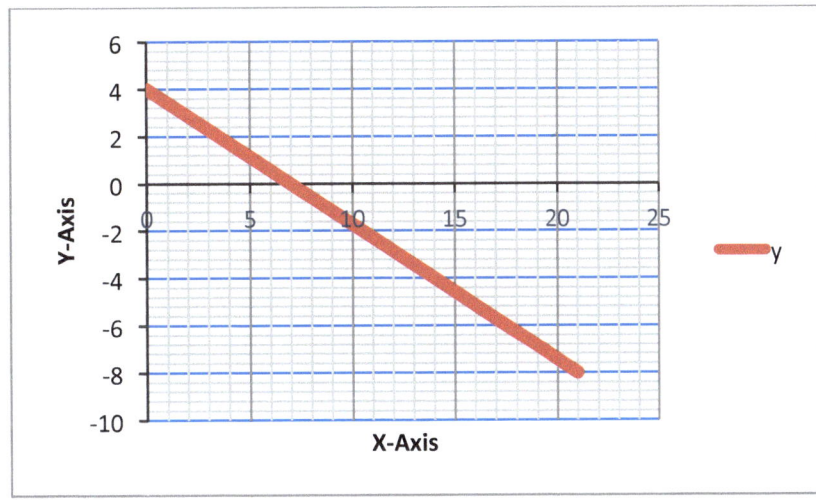

Figure 11

Maddy: Simple! With the rise (0 − 4) = -4 and the run (7 − 0) = 7, the slope equals $\frac{-4}{7}$.

Mentor: Can you describe this in words?

Maddy: Yes. For every seven units you go to the right, you will go down four.

Mentor: Excellent. Now here is our second slope calculation in Figure 12.

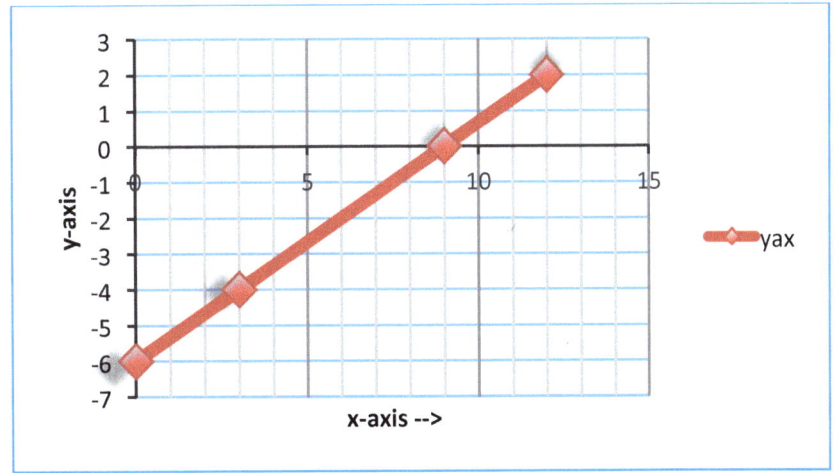

Figure 12

Emma: The slope is the rise, -6 − 0 = -6, divided by the run, 0 − 9 = -9, or -6/-9 = 2/3.

Mentor: You have managed well despite all the minus signs. Another approach would be to observe that from -6 on the y-axis up to the origin is six units and that from the origin to (9,0) on the x-axis is nine units, along with a visual observation that the rise is positive, gives us 2/3. You should be able to find the slope several ways so you can deal with a variety of situations. Now find the slope of the line in Figure 13.

15

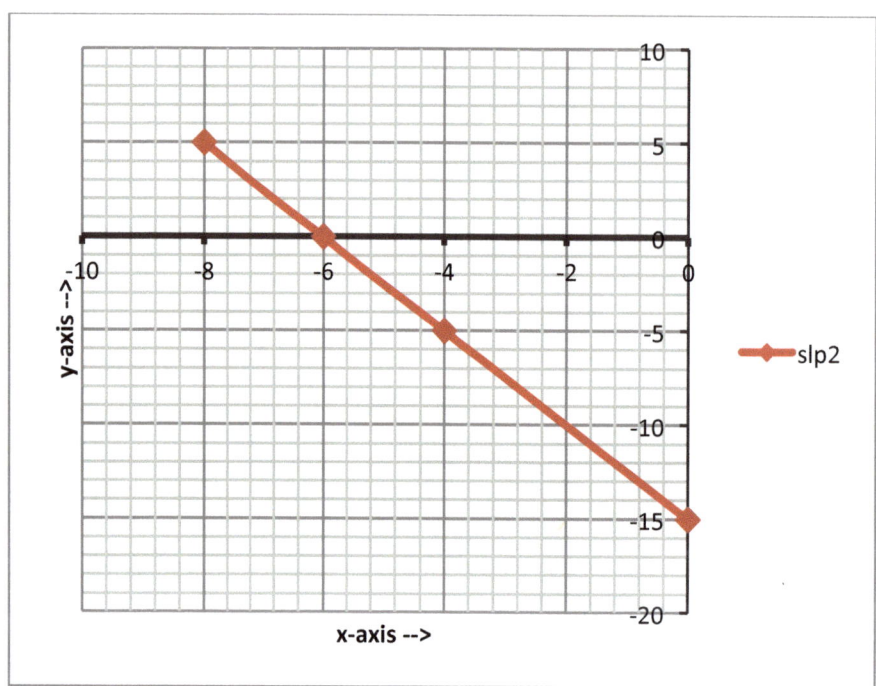

Figure 13

Ursula: It is getting pretty easy now. Seeing that the line drops 15 over a horizontal distance of 6 units, I know the slope equals -15/6 or -5/2.

Mentor: You have the idea. Now considering the equation y = 2x − 3, find the coordinates where this line intersects the axes.

Ollie: The line intersects the y-axis where x = 0, so y = -3. Also, the line intersects the x-axis where y = 0, so we have 0 = 2x − 3, making x = 3/2.

Mentor: Given those two points on the axes, calculate the slope of this line.

Finny: The slope is $\frac{0-(-3)}{\frac{3}{2}-0} = \frac{3}{3/2}$ or just 2. Now I see that the coefficient of x is the slope of the line!

Mentor: Our next scene will give you a chance to calculate slopes given any two points of the Cartesian plane.

Between the Scenes

What is the x-intercept and y-intercept of each of the given lines? What is the slope of each of the lines?

a) y = 4x − 7   b) y = -4x + 7   c) y = 5x + 2   d) y = -5x − 2

16

Act 1   **Scene 9**   **Slopes Given Any Two Points**

Mentor: In Scene 8, we were given the coordinates of a line where it passed through the two axes. We call these points the **intercepts**, the <u>x-intercept</u> where the line passes through the x-axis and the <u>y-intercept</u> where the line passes through the y-axis. In this scene, we will start with two points **not** on the axes and calculate the slope of the line. Here is a line passing through two points, and we want to know the slope of the line.

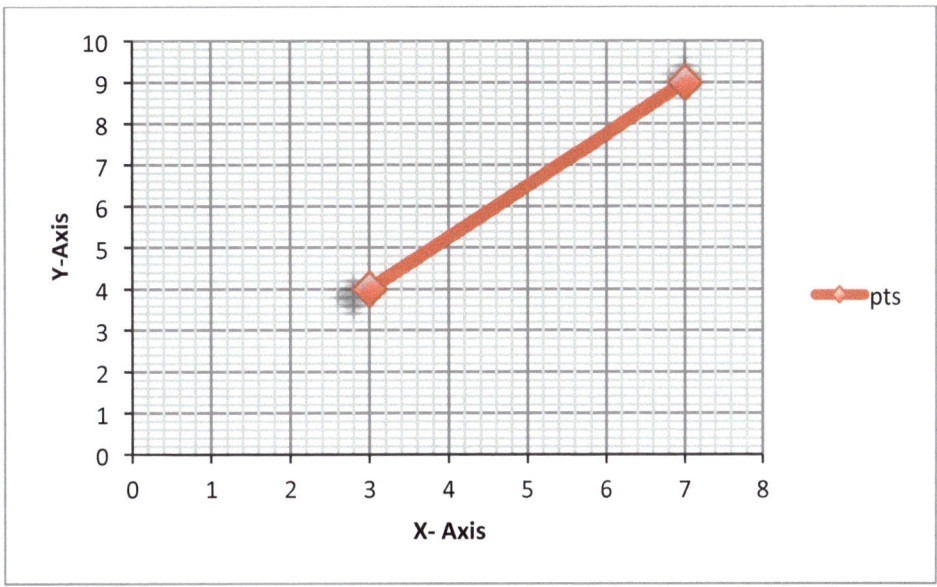

Figure 14

Piper: I know what to do. I see a triangle with two sides parallel to the x- and y-axes respectively making a third vertex at (7, 4). So the rise equals 9 – 4, and the run equals 7 – 3. So, we have the slope $\frac{9-4}{7-3} = \frac{5}{4}$.

Mentor: Piper makes that look pretty easy. Try using the two points A(-4, 9) and B(-8, 1).

Piper: That would be the same calculation $\frac{9-1}{-4-(-8)}$ or $\frac{8}{4} = 2$. So we can find the slope of a line with the given coordinates in a different quadrant, here quadrant II.

Mentor: Now let us try having the two points in two different quadrants as in Figure 15 on the next page.

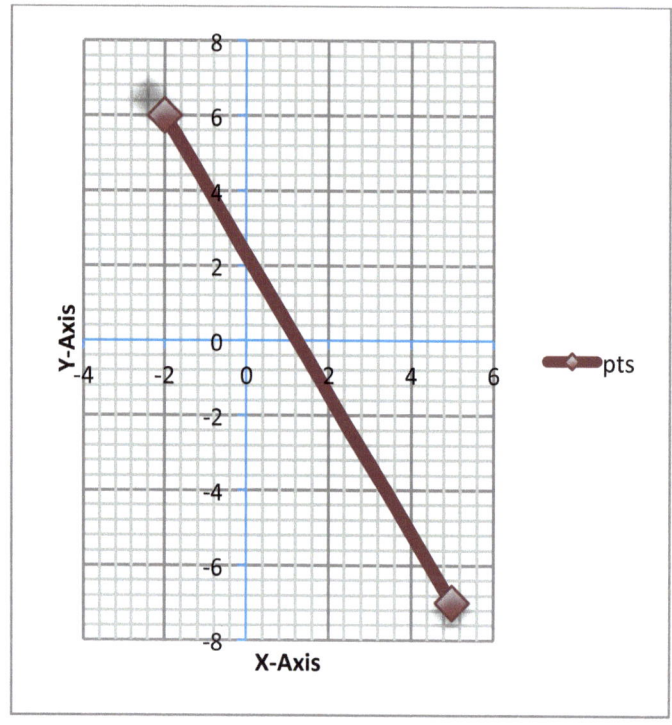

Figure 15

Gus: This is easy! I see a triangle under the line through the two points (-2, 6) and (5, -7). The triangle has its third vertex at the point (-2, -7), so each of the two sides not drawn is parallel to one of the axes. The slope is $\frac{6-(-7)}{-2-5} = \frac{13}{-7}$, or -13/7.

Mentor: Could you have calculated the slope a different way?

Gus: Yes. I could have subtracted the second quadrant coordinates from the fourth quadrant coordinates. Then the slope is $\frac{-7-6}{5-(-2)} = \frac{-13}{7}$ or -13/7 again.

Mentor: What do you think you can say in general about calculating the slope given two points?

Gus: We get the same slope if we subtract the corresponding coordinates of point A from point B as we would get if we subtracted the corresponding coordinates of point B from point A. Using <u>subscripts</u>, we could write: $\frac{y_A-y_B}{x_A-x_B} = slope.$

Mentor: Excellent. We note that we only subtract x-coordinates from x-coordinates and y-coordinates from y-coordinates. After an interlude solving linear equations we will be ready to find the equation of any line given the coordinates of two points on the line.

Between the Scenes

Calculate the slope of each of the following line segments and find their x- and y-intercepts.

i) PQ where P(-9, 2) and Q(1, -5)   ii) AB where $A(7,5)$ and $B(-\frac{1}{4}, \frac{1}{3})$.

iii) CD where $C\left(\frac{1}{6}, \frac{2}{9}\right)$ and $D(\frac{4}{15}, \frac{7}{10})$.

Act 1 Scene 10   **Opposite Operations**

Mentor: In the earlier parts of this act we have explored graphs of equations, linear equations such as $y = 7x - 2$. We discovered that solving one of these equations for x, such as $0 = 7x - 2$, involved using opposite operations from the ones in the equation. When do we perform the indicated operation and when, the opposite operation?

Piper: If I have to <u>solve</u> for x in $0 = 7x - 2$, using the opposite operations, I first add 2 to both sides.

$$0 = 7x - 2,$$
$$\text{adding 2,}$$
$$2 = 7x,$$
$$\text{then dividing by 7,}$$
$$\frac{2}{7} = x,$$
$$\text{or}$$
$$x = \frac{2}{7}$$

So, in solving the equation, we perform the opposite operation from the one indicated, and here we have done the operations in reverse order, namely addition before division or multiplication.

Mentor: Excellent. How does this differ from <u>evaluating</u> the expression for y if $x = 3$?

Trix: To find the value of y if $x = 3$, we substitute for x and calculate, using the indicated operations.

$$y = 7 \times 3 - 2,$$
$$\text{Multiplying by 7 before subtracting,}$$
$$y = 21 - 2,$$
$$\text{And completing the evaluation by subtracting,}$$
$$y = 19.$$

Mentor: Good!   Did you have something else you wanted to observe?

Trix: Yes. I noted that when Piper <u>solved</u> the equations, she did the operations in the <u>reverse order</u> from the usual multiplication and division before addition and subtraction.

Mentor: Another good observation. Could we have solved our equation in a different order?

Emma: Oh, yes. If I divided by 7 first and then add, I can calculate the solution again.

Mentor: Take us through the process!

Emma: Given: $0 = 7x - 2,$
$$\text{we divide both sides by 7,}$$
$$\frac{0}{7} = \frac{7x-2}{7},$$
$$\text{and dividing each term in the numerator by the denominator,}$$
$$0 = \frac{7x}{7} - \frac{2}{7},$$
$$\text{and simplifying,}$$
$$0 = x - \frac{2}{7}$$
$$\text{and adding we have,}$$
$$\frac{2}{7} = x, \text{ or } x = \frac{2}{7}$$

19

This is the same solution we had in the first process.

Mentor: Nice. So what are you observing about the order of operations in your solving process?

Emma: We can use the standard order (multiplication and division before addition and subtraction) but have to observe that we perform these operations on the whole side of the equation. Otherwise we will get the wrong solution.

Mentor: Once again, we find we can solve a problem, here an equation, in several ways. Does anybody have a third method?

Madeleine: Yes. I could <u>substitute</u> values for x until I hit the right one.

Mentor: Show us, please!

Madeleine: This could be quite lengthy, but with $0 = 7x - 2$, I might start by substituting $x = 0$ in red on Graph 1 below, because it is simple.

$$0 = 7 \times 0 - 2,$$
or
$$0 = -2$$

Because this equation is false, I look for another number, not negative, say $x = 1$ in blue.

$$0 = 7 \times 1 - 2,$$
or
$$0 = 5$$

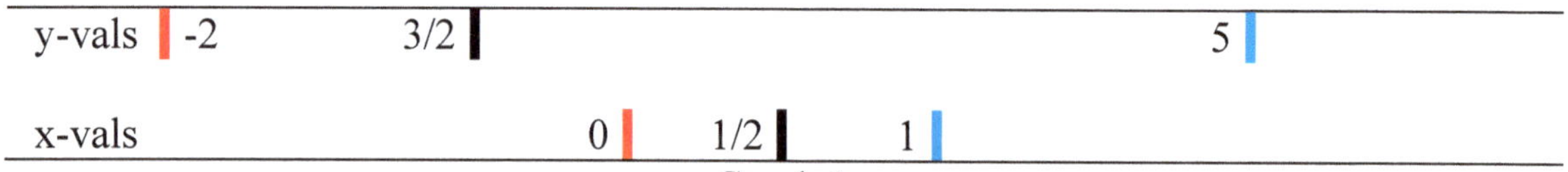

y-vals	-2	3/2		5

x-vals		0	1/2	1

Graph 1

Because we have found two values, one above 0 and one below, we search in between, with say $x = \frac{1}{2}$ in black.

$$0 = 7 \times \frac{1}{2} - 2,$$
or
$$0 = \frac{7}{2} - \frac{4}{2},$$
or
$$0 = \frac{3}{2}$$

Again we have a positive number on the right side, so we pick a smaller number between 0 and $\frac{1}{2}$, say $\frac{1}{4}$, shown in black on Graph 2 on the next page.

Mentor: This IS quite lengthy!

Madeleine: Yes, but I am getting closer.

$$0 = 7 \times \frac{1}{4} - 2,$$
making a common denominator,
$$0 = \frac{7}{4} - \frac{8}{4},$$
or
$$0 = -\frac{1}{4}$$

20

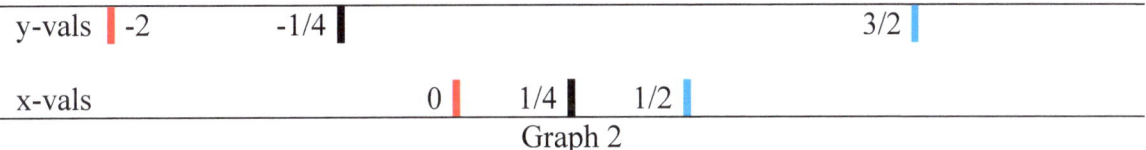

y-vals	-2	-1/4		3/2

x-vals		0	1/4	1/2

Graph 2

While this equation is false, we do have a value on the right closer to zero as seen in Graph 2. This time we pick a value between $\frac{1}{4}$ and $\frac{1}{2}$, because $-\frac{1}{4} < 0 < \frac{3}{2}$.

This time I pick a value that might simplify my arithmetic, such as $\frac{3}{7}$.

$$0 = 7 \times \frac{3}{7} - 2,$$
or
$$0 = 3 - 2,$$
or
$$0 = 1$$

y-vals	-1/4	1	3/2

x-vals	1/4	3/7	1/2

Graph 3

Now $\frac{3}{7}$ is too big, shown in black on Graph 3, so we will try $\frac{2}{7}$ and see that it works!

$$7 \times \frac{2}{7} - 2 = 2 - 2, \text{ or } 0.$$

Mentor: Your persistence paid off finally. This approximation process can take many steps, but it does provide us with yet another way to find a solution.

Between the Scenes
Solve each of the equations for x in at least two different ways.
a) $0 = -5x + 17$ b) $0 = 4x - 17$ c) $0 = -7x + 6$ d) $0 = 11x + 8$

Act 1   Scene 11   Graphing and Equations

Mentor: In Scenes 3 and 5, we have looked at two kinds of changes we can make to the equation y = x and can see that each change has its own distinctive effect. If we combine multiplication and subtraction as in the equation $y = 4x - 3$, what can we say about its graph?

Madeleine: The 4 does a rotation-like twist on the line and the -3 drops the line down three units on the y-axis.

Mentor: Perhaps we should attempt to describe the "rotation-like twist" a little more accurately. On any given vertical line, say x = 2, the point (2, 2) on y = x will be moved upward away from the x-axis by a multiple of 4, to the point (2, 8). Note: an upward movement. A point below the x-axis, say (-2, -2), will be moved down along the line x = -2 to (-2, -8), four times the distance of (-2, -2) from the x-axis.

Please show us what those two lines, y = 4x and $y = 4x - 3$, look like.

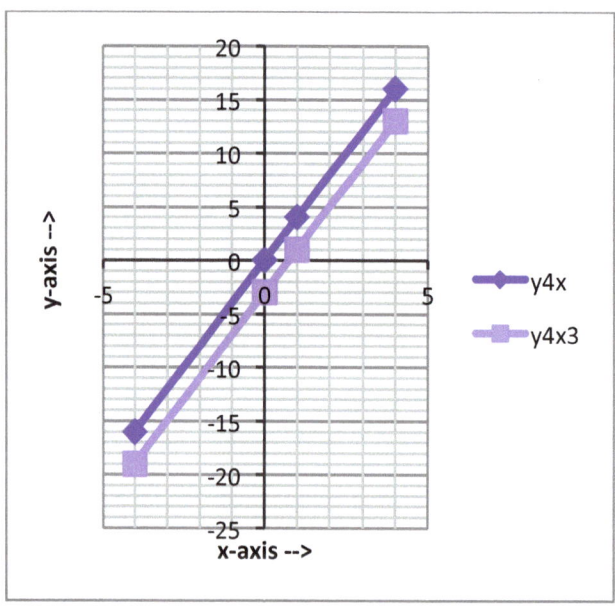

Figure 16

Madeleine: The dark purple (diamond) line segment represents y = 4x and the light purple segment shows the corresponding points on y = 4x – 3. We see that the light purple line crosses the y-axis at (0, -3), three units below the dark purple line.

Mentor: We call this point the y-intercept, where the light purple line intercepts the y-axis. What is the value of the x-intercept, where this line intersects the x-axis?

Georgia: The x-intercept is between 0 and 1, because the y-intercept is -3 and the y-coordinate at x = 1 is 1.

Mentor: Excellent. If a line has a negative y-coordinate at one point and a positive one at another, then it must intersect the x-axis in between the two points. Notice that on our graph the scales on the x- and y-axes are not the same. How can we find the coordinate of the x-intercept?

Georgia: Let's draw a graph of just that part of the line between 0 and 1.

22

Mentor: Excellent. What points do you want to show?

Georgia: I want to take x-coordinates $\frac{1}{4}$, and $\frac{1}{2}$, and $\frac{3}{4}$, because multiplying each by 4 will result in an integer. So, we have ($\frac{1}{4}$, -2), ($\frac{1}{2}$, -1), and ($\frac{3}{4}$, 0). We got it!

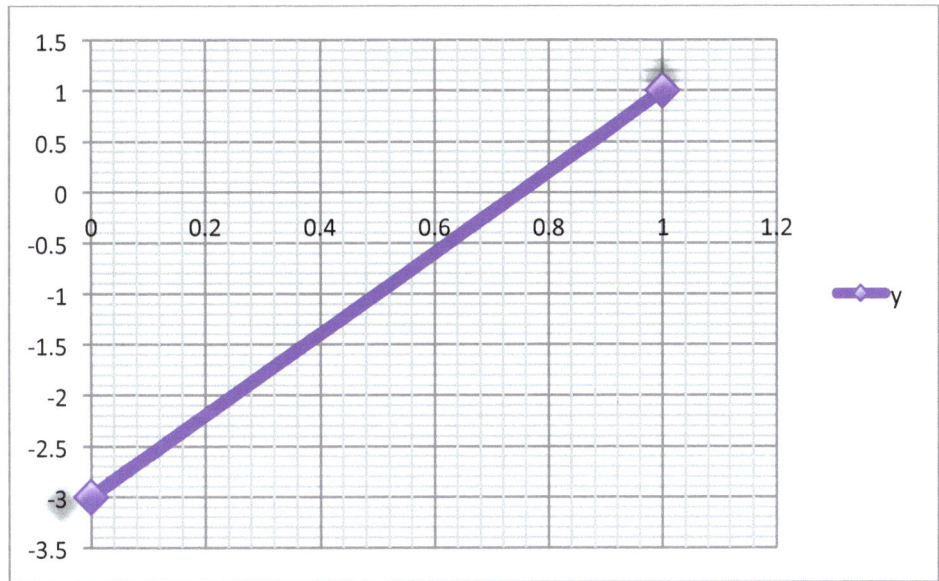

Figure 17

Mentor: Graphing actually led you to calculate the exact value. This will not always work, but it makes a good start. Suppose we wanted to know what value of x would give us y = 0 in the equation y = 5x + 2. Have we seen how we might proceed?

Georgia: Yes. I would pick x equal to some fraction with 5 in the denominator, so the result of multiplying 5x would be an integer.

Mentor: And what integer do you want the result to be?

Georgia: Negative two! So, I will pick $x = \frac{-2}{5}$.

Mentor: Now we have the x-intercept where y = 0, what would the y-intercept be?

Georgia: The constant term, here 2.

Mentor: Let's try another example. What are the x- and y-intercepts of the line y = -3x + 7?

Gus: If x = 0, then y = 7. So, 7 equals the y-intercept. To make y = 0, I have to divide -3 by 3 for -1 and multiply it by 7. Giving us -3(7/3) + 7 = 0. The x-intercept equals $\frac{7}{3}$.

Mentor: I guess all of you have learned how to calculate the fractional parts of a number! Does anyone have a different approach?

Piper: Yes. I would write the equation 0 = -3x + 7, and add 3x to both sides, giving me 3x = 7. To get just x on the left side I have to divide by 3 on both sides.

Mentor: Piper has just started us off on the concept of solving a "linear" equation. This will be the subject of our next scene.

Act 1   Scene 12   X- and Y-Intercepts

Mentor: Let us suppose we are given two points, not on either axis, and want to find the x- and y-intercepts of the line joining them. Consider the line through the points A(-4, -8) and B(-9, -2). What are its x- and y-intercepts?

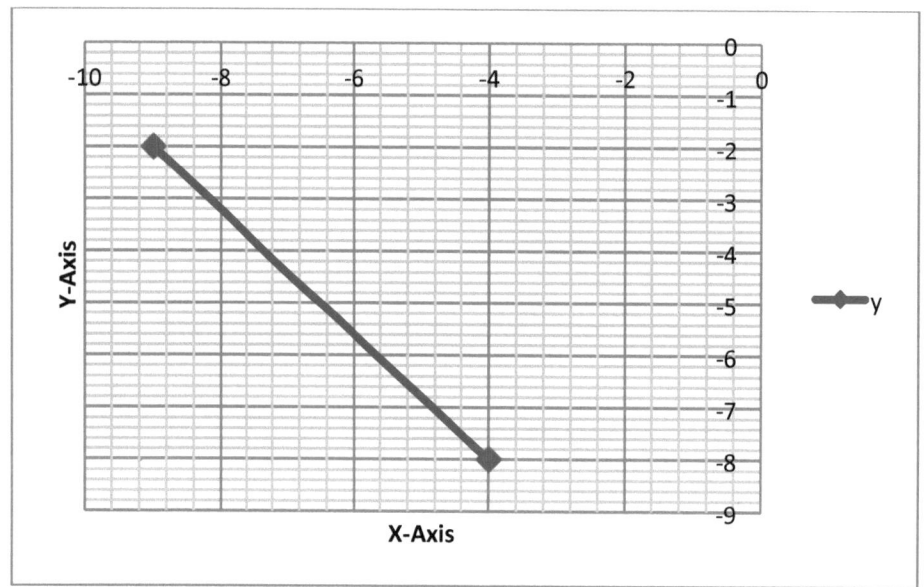

Figure 18

Matthew: I want to calculate the slope first, then the y-intercept and finally the x-intercept. The <u>slope equals the rise over the run</u>, or slope $= \frac{-2-(-8)}{-9-(-4)} = \frac{-2+8}{-9+4}$, or $\frac{6}{-5} = -\frac{6}{5}$.

Mentor: What comes next?

Matthew: I can write an equation, filling in the slope times the x-coordinates. So, now we have: $y = -\frac{6}{5}x + ?$.

Mentor: You are doing well, but let me introduce some standard notation. When we write the equation of a line without knowing the numerical values of the slope and y-intercept, we fill in "<u>m</u>" <u>for the slope</u> and "<u>b</u>" <u>for the y-intercept</u>. In this way we do not have to use question marks or other symbols. This way we can write "m $= -\frac{6}{5}$" and everybody knows m is the slope of this line. We also use the letter "b" for the y-intercept.

Matthew: Then I have the equation $y = -\frac{6}{5}x + b$. Now I do not know what to do!

Mentor: Do you know a point this line must go through?

Matthew: Sure. (-9, -2). Oh. I can replace x by -9 and y by -2, and then solve for b!

$$-2 = -\frac{6}{5} \times (-9) + b$$

Mentor: How do you propose to do that?

24

Matthew: First I simplify the product of two negative numbers, giving us the equation:

$$-2 = \frac{6}{5} \times 9 + b$$

Then I am going to multiply the fraction times 9 to simplify the expression.

$$-2 = \frac{54}{5} + b$$

Seeing that b + 54/5 equals -2, I am going to subtract the fraction from -2.

$$-2 - \frac{54}{5} = b$$

This gives us b = -64/5.

Mentor: What is the equation now?

Matthew: The equation is: $\qquad y = -\frac{6}{5}x - \frac{64}{5}$

And we know that the y-intercept is $-\frac{64}{5}$.

Now we have to solve the equation: $\quad 0 = -\frac{6}{5}x - \frac{64}{5}$. If I multiply by 5, the new
equation is: $\qquad\qquad\qquad 0 = -6x - 64$.

Adding 6x to each side gives us: $\quad 6x = -64$.

Dividing by 6: $\qquad\qquad\qquad\quad x = -64/6 \text{ or } -32/3$.

Mentor: That was a good effort with so many fractions. How could we have avoided using so many fractions?

Matthew: We could have multiplied by 5 first to give us 0 = -6x – 64, then added 6x, resulting in the equation 6x = -64. This would have been easier!

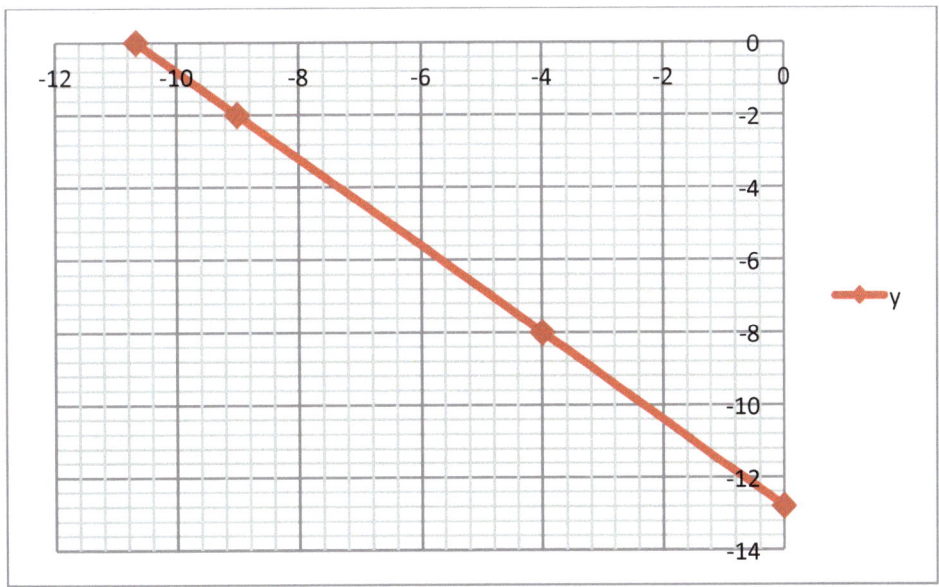

Figure 19

Act 1   **Scene 13**   **Line Equation Writing Without Calculation**

Mentor: We have seen how much computation goes into taking the coordinates of two points and writing the equation of the line passing through them. In this scene we will investigate a way of writing an equation of a line without any calculations. If we were given three points (A, B, and C) on a line, what could you say about the slopes of the line segments AB, AC, and BC?

Gus: The slope of a line should be the same taken from any point to any other point.

Mentor: So, if I considered the two points, A(-4, -8) and B(-9, -2), from the previous scene and a third point, C(x, y), what do you think you could say about the slopes of the segments AB, BC and AC?

Gus: They are all equal!

Mentor: Can you give us an equation involving the slopes of segments AB and AC?

Gus: The simple part repeats the previous scene and adds a second slope.

$$\frac{y - (-8)}{x - (-4)} = \frac{-2 - (-8)}{-9 - (-4)}$$

Mentor: If we want to use this equation to calculate the x- and y-intercepts, what would we perhaps want to do to simplify our work?

Gus: We should simplify the fractions with all the minus signs and negative values. We know from Scene 12 that the slope equals -6/5. Also, we can write the left side without all the parentheses and negative numbers.

$$\frac{y + 8}{x + 4} = -\frac{6}{5}$$

Mentor: How could we find the x- and y-intercepts using this equation?

Ursula: For the x-intercept, I would replace y by 0 and solve for x. Similarly, for finding the y-intercept, replacing x by 0.

Mentor: Please take us through the finding of the x-intercept.

Ursula: Setting y equal to 0, we have:

$$\frac{8}{x + 4} = -\frac{6}{5}$$

If I multiply both sides by 5 and by x + 4, I will have:
$$40 = -6(x + 4)$$
Dividing both sides by 2: $20 = -3(x + 4)$

Now divide both sides by -3: $\frac{20}{-3} = x + 4$

Subtracting the 4 and simplifying: $-\frac{20}{3} - 4 = x$, or x = -32/3

Mentor: Excellent. This is, of course, the same value we got in the previous scene. Each of you should try to calculate the y-intercept in the same manner.

Act 1   Scene 14   Intersecting Lines

Mentor: We have looked at the equations of lines in a number of different ways, so you should be able to tell me when two lines intersect.

Matthew: We have seen in Scene 3 that two lines with the same slope are parallel, or non-intersecting, so I am guessing that lines with different slopes intersect.

Mentor: Sounds good. Suppose we look at the two lines below. Tell us where they intersect.

$$y = 4x - 5$$
$$y = -3x + 1$$

Matthew: They intersect where the x- and y-coordinates are equal. So from y = y we get:

$$4x - 5 = -3x + 1$$

Now I will try to get all the x's on the left side and the constants on the right, first by adding 3x to both sides and then adding 5 to both sides.

$$7x - 5 = 1$$
and
$$7x = 6$$

Now I have to divide both sides by 7 to have just an x on the left.

$$x = \frac{6}{7}$$

Mentor: You have done the hard part, but what is the y-coordinate at the intersection?

Matthew: I have to substitute 6/7 into one of the given equations.

$$y = 4 \times \frac{6}{7} - 5$$
or
$$y = \frac{24}{7} - \frac{35}{7}$$
and
$$y = -\frac{11}{7}$$

Mentor: So where is this point?

Matthew: The point $(\frac{6}{7}, -\frac{11}{7})$ is in quadrant IV. Here is the graph of the two lines. We can see from the graph that the x-coordinate is close to 1 and the y-coordinate is a little less or lower than -1.

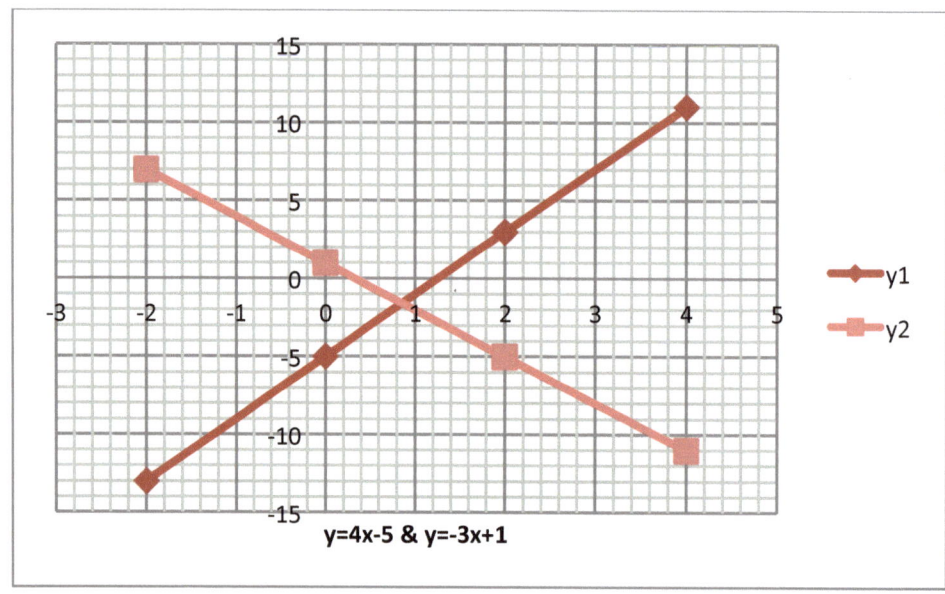

y=4x-5 & y=-3x+1

Figure 20

Mentor: Can anybody describe the process we just went through?
Piper: Yes. First, we set the y-coordinates equal, using the right side of each equation. Second, we added an x-term to make the x appear only on one side. Third, we added a constant term to remove the constant from the side where the x-term is. Fourth, we multiplied or divided by a constant to change the coefficient of the x-term to 1.
Mentor: Good description! Suppose I gave you two points on each of two lines and asked you to find the point of intersection of the two lines. What would you do?
Ursula: I would find the equations of each of the lines and then find the intersection as we just did here.
Mentor: I'm thinking of two lines, AB and CD, with the coordinates A(-8,3) and B(1, -15); with C(1, 15) and D(-6, -6), and I want to know where the two lines intersect. Who has the slope of AB?
Emma: The slope of AB is $\frac{3-(-15)}{-8-1} = \frac{18}{-9}$ or -2.
Mentor: And the y-intercept?
Emma: The equation for AB now is: y = -2x + b. Replacing x and y with the coordinates of A, we have 3 = -2(-8) + b, giving us b = 3 – 16 or -13. The equation for line AB: $y = -2x - 13$.
Ursula: And I have the equation of CD as: y = 3x + 12.
Piper: Next we solve the equation: $-2x - 13 = 3x + 12$. Adding 2x and subtracting 12, we have -25 = 5x, or x = -5. Now we have y = 3(-5) + 12, or y = -3. The intersection is
(-5,-3).
Mentor: You all have the idea. From Figure 21 (on the next page) we can see that our solution is about right.

28

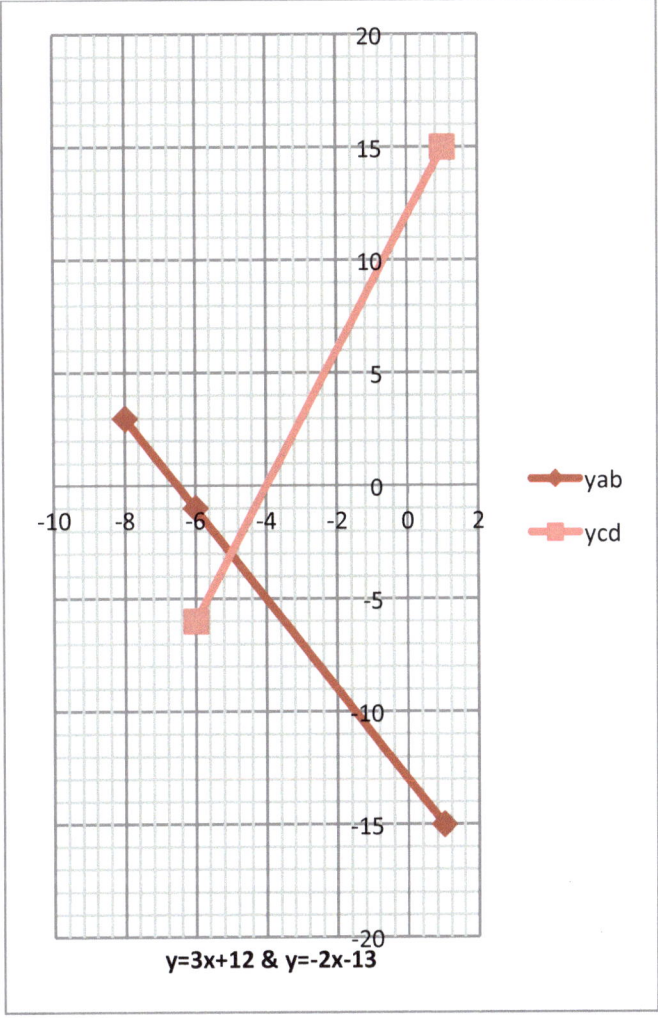

y=3x+12 & y=-2x-13

Figure 21

Mentor: We also might point out that we could have drawn this picture <u>before</u> making the calculations, so we knew about where the solution would be. A quick picture may also help you catch any mistake.

Between the Scenes
1) Find the intersections of the following pairs of lines.
    a) $y = -6x - 7$ and $y = -8x + 9$
    b) $y = 4x + 11$ and $y = -5x + 2$
    c) $3x + 5y = 17$ and $-2x + 7y = -5$
    d) $y = 2x - 7$ and $y = -3x + 1$
2) Alarm-repairman David's problem-solving method:
    i) Every problem comes in multiple parts.
    ii) Ask the customer (ie the problem) what has changed.
    iii) Cut the problem in half.
    iv) Make things smaller.
    v) Use visual and verbal representations

Act 2   Visualization and Generalization

Scene 1   Summing the first 10 Cuisenaire Rods

Mentor: In the first lessons of our arithmetic class, most students not only made a staircase out of the ten rods but also made a rectangle by putting a second set of stairs on top of the first in reverse order. How did we calculate the area of the stairs?

Figure 22

Madeleine: First we recognized that the area of a rectangle equals its length times its width, here 10 across by 11 units high, or 110 cm². Next, we recognized that we had two sets of stairs and then had to divide the rectangle's area by two, giving us 55 cm².

Mentor: So, now we have two places to use letters instead of the particular numbers. How can we describe the area of a rectangle?

Madeleine: The area of a rectangle equals the product of the length times the width.

Mentor: Yes. If we wanted to use just one letter to represent the "area of a rectangle," what letter would you use?

Madeleine: A.

Mentor: Similarly, what one letter would you use for the length and another for the width?

Madeleine: L for the length and W for the width.

Mentor: What equation would express the relationship between these three quantities?

Madeleine: $A = L \times W$.

Mentor: In our situation, we have a special case involving the number of Cuisenaire Rods we are using to make the steps, or 10. Suppose I used the letter "n" for the number of steps. How would we rewrite the formula for the area of the rectangle?

Ursula: The length across the bottom of the rectangle would be "n" and the width or height would be "n + 1." So our rectangle's area would be: $A = n(n + 1)$ and our stair's area would be $A = \frac{n \times (n+1)}{2}$, or spelled out: $1 + 2 + \ldots + n = \frac{n(n+1)}{2}$.

Mentor: We could use subscripts, r for rectangle and s for stairs, giving $A_r = n(n + 1)$ and $A_s = \frac{n(n+1)}{2}$, respectively. {In sigma form: $\sum_{i=1}^{n} i = \frac{n(n+1)}{2}$ means the same as $1 + 2 + \ldots + n = \frac{n(n+1)}{2}$, where the variable i takes on all the values from 1 to n.}

Act 2   Scene 2   The Difference of Two Squares

Mentor: Another calculation we have done using only numbers involves the difference of two squares. Let us consider the pattern for the factored form with numbers first. How can we factor $9^2 - 5^2$?

Ursula: $9^2 - 5^2 = (9 - 5)(9 + 5)$.

Mentor: Correct. Now let us write this arithmetic formula as an algebraic formula, using letters where there are numbers.

Ursula: $x^2 - y^2 = (x - y)(x + y)$.

Mentor: You know how to do this without using a picture, but please show us what you had in your mind's eye when factoring this difference.

Ursula: I picture x as the larger number and y as the smaller number, and I place the smaller square in the lower right-hand side of the larger square.

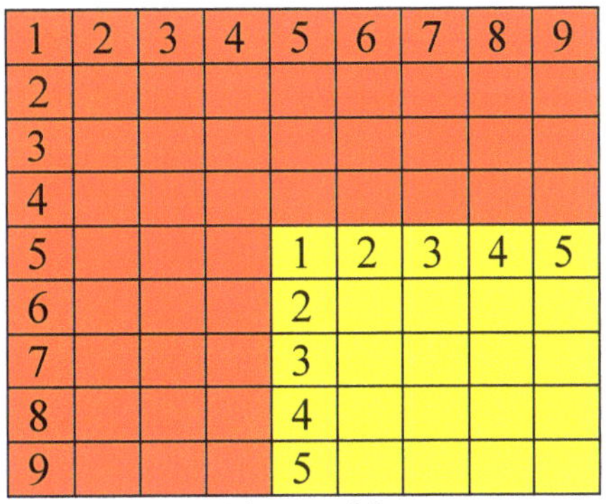

1	2	3	4	5	6	7	8	9
2								
3								
4								
5				1	2	3	4	5
6				2				
7				3				
8				4				
9				5				

Figure 23

Ursula: Then I remove the smaller square from the picture, leaving me with an upside down "L."

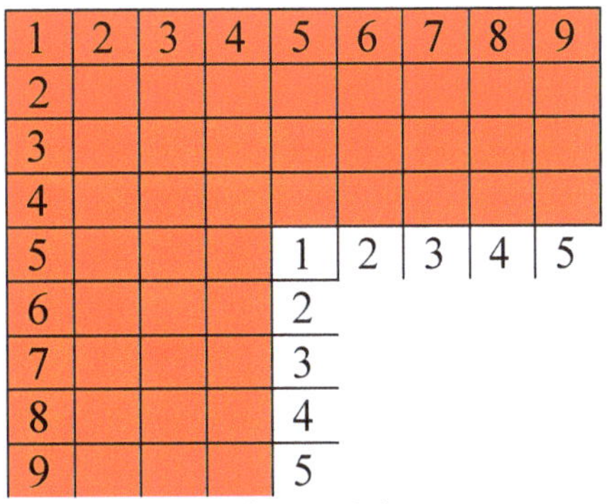

1	2	3	4	5	6	7	8	9
2								
3								
4								
5				1	2	3	4	5
6				2				
7				3				
8				4				
9				5				

Figure 24

Ursula: Now we chop off the part of the "L" that sticks out beyond the vertical section, leaving us a rectangle $x - y$, or here $9 - 5$, units across the top and 9 units high. We will swing the cut-off around so that its x, or here 4 units side, fits on the bottom of the remaining vertical section. This will give us a new rectangle that is $x - y$ across the top and $x + y$, or here $9 + 5$, down the side, with an area of $(x - y)(x + y)$.

31

Figure 25

Mentor: So we see that removing one square from another leaves matching sides, one four units long on the horizontal bottom side and the other four units long on the vertical right side of the upside down "L" in Figure 24.

Between the Scenes
Factor the following differences:
a) $9x^2 - 4y^2$　　　b) $y^2 - 16x^2$　　　c) $x^4 - y^4$　　　d) $x^6 - y^6$

Complete the following partially factored expressions:
e) $x^3 - y^3 = (x - y)(? + ? + ?)$　　　f) $x^3 + y^3 = (x + y)(? - ? + ?)$

g) $8x^3 - 27y^3 = (2x - 3y)(? + ? + ?)$　　　h) $x^3 + 125y^3 = (x + 5y)(? - ? + ?)$

i) $y^6 - x^6 = (y - x)(? + ? + ?)(y + x)(? - ? + ?)$

Act 2   Scene 3   The Product of Consecutive Odd Integers

Mentor: The product of consecutive odd numbers gives us another example of learning something about numbers through looking at an illustrating model as well as experiencing a way of expressing odd numbers in a general manner. Name two consecutive odd numbers and calculate their product.

Trix: 5 and 7 are consecutive odd numbers with a product of 35.

Mentor: State another pair and their product.

Trix: 7 and 9 with the product 63.

Mentor: Are these products related in any way to the number between the two consecutive odd numbers?

Trix: Well, $6^2 = 36$ and $8^2 = 64$. Let me try 9 and 11. Yes. 9 times 11 equals 99 and the square of 10 equals 100. It would seem to be that the product of two consecutive odd numbers equals one less than the square of the even number between them.

Mentor: Excellent. How could you show me visually that this is true?

Trix: If I took 5 rows of 7 white rods, I could take the last column of 5 rods and put them along the top, almost making a square 6 on a side. See Figure 26 below.

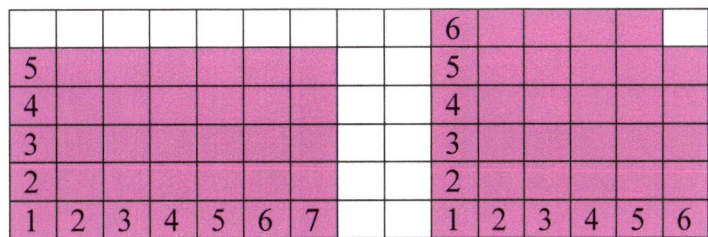

Figure 26

Mentor: Do you see this argument or demonstration working for all pairs of consecutive odd numbers? Why?

Trix: For larger numbers, I see Figure 26 as showing just the upper right-hand corner of the whole rectangle and its nearly-a-square changed form. The smaller odd number will never complete the longer row as it is two rods short from the start.

Mentor: Let's consider how we can represent this product with letters, such as n. How does an even number look?

Gus: We could write an even number as 2n.

Mentor: And the odd numbers on either side?

Gus: $2n - 1$ and $2n + 1$. Oh, they look like the two factors in the difference of two squares! We know that $(2n - 1)(2n + 1) = (2n)^2 - 1^2$ or $4n^2 - 1$.

Mentor: Does Gus's argument depend on the value we choose for n?

Class: No. The argument works for all natural numbers! It's a pattern.

(Prof. Edward B. Burger's <u>An Introduction to Number Theory</u> suggested this pattern, using blocks!)

Act 2　　　Scene 4　　　Summing the First n Odd Numbers

Mentor: Now that we have learned how to express odd numbers in a general form, we can consider the sum of the first n odd numbers. How can we express the nth odd number?

Georgia: How about using 2n + 1?

Mentor: A good idea, but how do we express 1?

Georgia: I guess n would have to be equal to 0?

Mentor: That would work, but how would you arrive at the third odd number, 5?

Georgia: I would have to use n = 2, so 2n + 1 would be equal to 5.

Mentor: Do you see a way to represent the third odd number using the number 3 instead of 2?

Georgia: Oh! I see that I should have chosen the formula: 2n – 1.

Mentor: Now let's list the first 10 odd numbers and look for a pattern of sums.

Georgia: 1, 3, 5, 7, 9, 11, 13, 15, 17, 19. If we add 1 + 3 we get 4. If we add 5 to that we get 9, and adding 7 we are up to 16. We keep getting squares!

Mentor: Let's create a table to see this pattern.

Counting: ↑

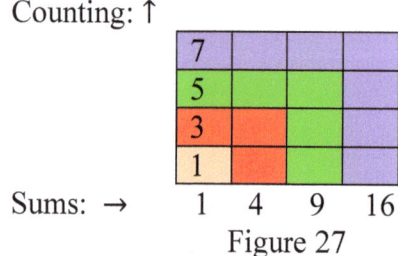

Sums: →　　1　4　9　16

Figure 27

Georgia: I see the pattern now! The sum of the first n odd numbers equals n^2. Every "L" shaped section we put on the square makes another square!

Mentor: That's good observation. {For those of you who like the details, we can again use the sigma notation to describe this sum as follows: $\sum_{t=1}^{n}(2t - 1) = n^2$. We start counting with t = 1 and end when t = n. The expression 2t – 1 changes as we replace t by its current value: $2\times1 - 1 = 1$, and $2\times2 - 1 = 3$, and $2\times3 - 1 = 5$, and so on up the left side of our Figure 27.}

Between the Scenes

Sum in two different ways the first 10 natural numbers having a remainder of 1 after being divided by 3: $\sum_{i=1}^{i=10}(3i - 2)$.

　a) 1 + 4 + 7 + … + 28 = ?

　b) Add the columns

$$3 \times 1 - 2$$
$$3 \times 2 - 2$$
$$3 \times 3 - 2$$
$$…$$
$$3 \times 10 - 2$$

　c) Now find the sum of the first 100 natural numbers having a remainder of 1 when divided by 3.

34

Act 2   Scene 5   Pythagoras Through Pictures

Mentor: In the third lecture in his <u>Philosopher's Toolkit</u>, Professor Patrick Grim suggests that Pythagoras did not have the proof of the theorem named after him that we all learn in school. That proof was the product of Euclid some time later. Let us consider creating a visual "proof." But first we must state this useful theorem.

Emma: I know it! The sum of the squares on the legs equals the square on the hypotenuse.

Mentor: You have it. Let us make that statement visual. First we draw a 3-4-5 right triangle, or any other right triangle, shown in Figure 28. Then what will we do?

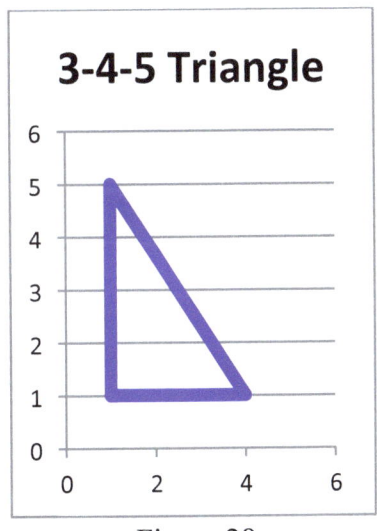

Figure 28

Emma: We will draw squares on each of the sides of the triangle.

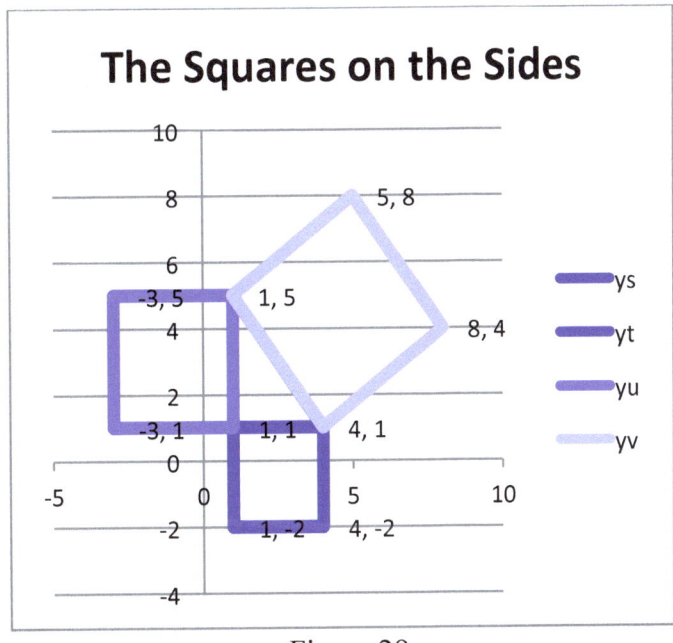

Figure 29

35

Mentor: And now what? How can we show that the area covered by the biggest square is the same as the area covered by the two smaller squares? What would your strategy be? I needed some help from Grim. So, if you want hints, read on.

Hint 1. Make three copies of your triangle, so you have four to work with.

Hint 2. Try to cover up as much of the largest square with the four triangles as you can.

Hint 3. Can you arrange the triangles so that you have an uncovered square in the middle of the largest square? Create a square exactly the size of the uncovered area.

Hint 4. Take your four triangles and little square to see if they can cover the combined area of the two smaller squares.

Hint 5.The lengths of the sides of the triangles should match the sides of the square or squares to be covered.

If you want to see a representation of a solution, look on the next page.

Solution to the visual proof of the Pythagorean Theorem.

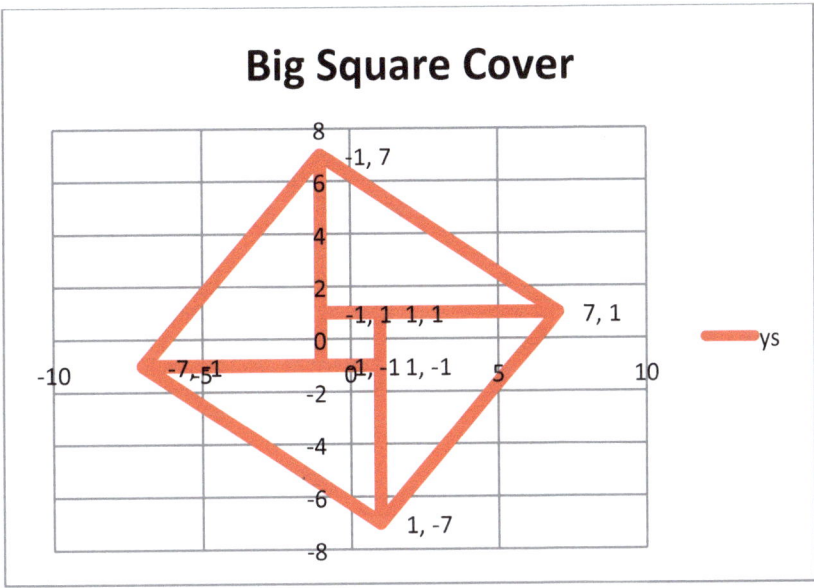

Figure 30

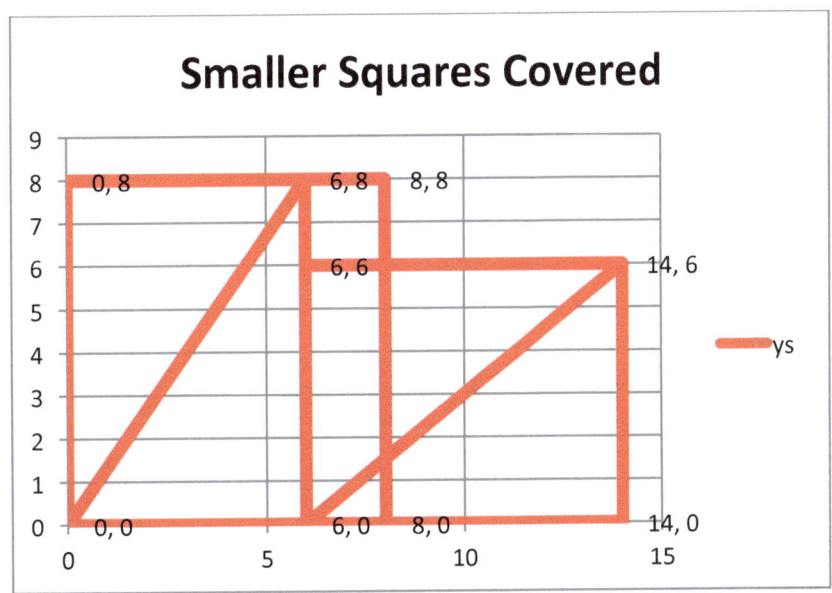

Figure 31

Act 2   Scene 6   Distances in the Cartesian Plane

Mentor: In Scene 5 we have offered an informal proof that the square of the hypotenuse of a right triangle equals the sum of the squares of the legs. Given the right triangle below, how can we calculate the length of each of the three sides?

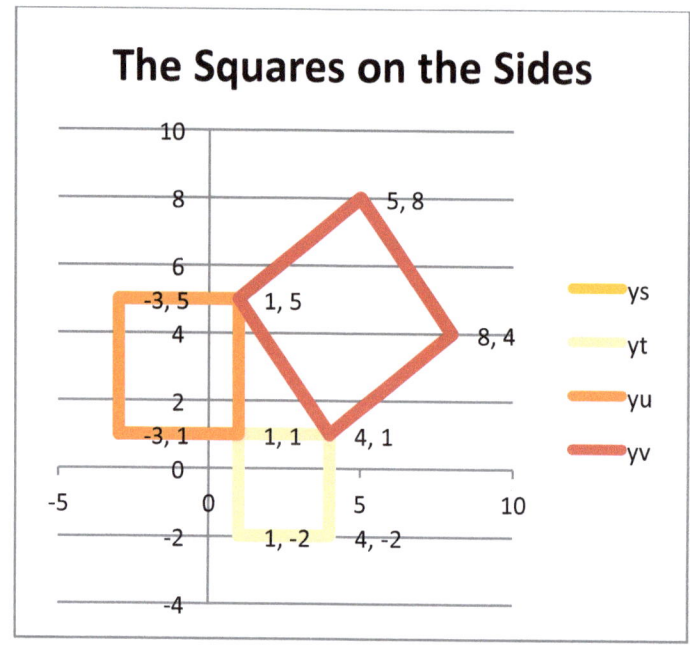

Figure 29B

Emma: I will start with the right side of the small square at the bottom. I see that the x-coordinates of the two points on the same vertical line, (4, 1) and (4, -2), are the same, indicating no change in horizontal distance. Then I look at the y-coordinates and see that the difference of the coordinates, subtracting smaller from larger, gives the vertical distance, or length, along that line: $1 - (-2) = 3$. This system works for both the smallest square at the bottom and the square on the left side: $5 - 1 = 4$.

Mentor: This works well for vertical line segments and for all four sides of a square, but how would you calculate the length of the horizontal sides starting from scratch?

Emma: I would just use the x-coordinates on the horizontal segments. On the bottom triangle side, we find (1, -2) and (4, -2), indicating no change of vertical distance. So I would subtract the smaller x-coordinate, 1, from the larger, 4, to arrive at 3 units long: $4 - 1 = 3$. For the triangle on the side we subtract the smaller y-coordinate, 1, from the larger, 5, to get 4.

Mentor: You have given us good directions for calculating the lengths of horizontal and vertical segments in a Cartesian plane, but what are we to do with the biggest square on the hypotenuse?

Trix: The Pythagorean theorem we have just "proved" tells us that we can use the lengths of the sides of the squares on the legs to calculate the size of the square on the hypotenuse:

$$h^2 = (5 - 1)^2 + (1 - (-2))^2, \text{ or } h^2 = 4^2 + 3^2.$$

Mentor: Excellent! But what must we do to find the length of a side on the hypotenuse?

Trix: Now I have to take the square root of the $3^2 + 4^2$ expression, or $\sqrt{3^2 + 4^2} = h$. Here we see that the hypotenuse equals 5.

Mentor: You have done well to travel a long path. Could you write one expression that would indicate all the operations you had to perform in order to calculate h?

Trix: I think so. I will go back to the original expression for the length of h, where the square of h equals the sum of the squares of the legs of the right triangle:

$$h^2 = (5 - 1)^2 + (1 - (-2))^2$$

Now all we have to do is to take the square root of the right side to calculate the length of h.

$$h = \sqrt{(4 - 1)^2 + (1 - (-2))^2}$$

So the hypotenuse equals the square root of the sum of the squares of the lengths of the legs of the right triangle. And we can read the lengths of the legs, assuming one is horizontal and the other vertical, by subtracting the smaller coordinate from the larger, called the <u>absolute value</u> of the difference, and square that difference.

Mentor: You have brought this section to a close.

Between the Scenes
Calculate the lengths of the following segments.

1. AB where A = (7, 5) and B = (9, -2).
2. CD where C = (4, 6) and D = (-7, 6).
3. EF where E = (7, 1) and F = (7, -4).
4. GH where G = (-11, 3) and H = (11, -5).

Act 2   Scene 7   A Reflecting Telescope

Mentor: Does anybody know how a reflecting telescope works?
Madeleine: Sure. It takes rays of light from a long way away and focuses them on one point, building up a brighter light than any one ray.
Mentor: What do we know about all those rays from a star millions of miles away?
Madeleine: They are almost exactly parallel to each other.
Mentor: Could we use a flat mirror to gather this light?
Madeleine: We could use a flat mirror to turn all the rays of light around, but they would remain parallel to each other, unfocused.
Mentor: What do we want to do to each of these incoming rays?
Madeleine: We want to bend all of them into a single point.
Mentor: Suppose we think of the mirror being one unit below the x-axis and the point of focus on the y-axis one unit above the origin. Can you draw this setup?
Madeleine: The mirror has an equation, $y = -1$, and the focus point has coordinates F(0, 1).
Mentor: If we wanted the parallel rays of light to hit the focus point, F, at the same time, how far should all the rays travel?
Madeleine: So the distance from each reflection point, P(x, y) to the flat mirror at G, should be the same as the distance from the focus point, F, requiring a curved mirror.
Mentor: I'm not sure I can see what you are saying.
Madeleine: Here is my picture.

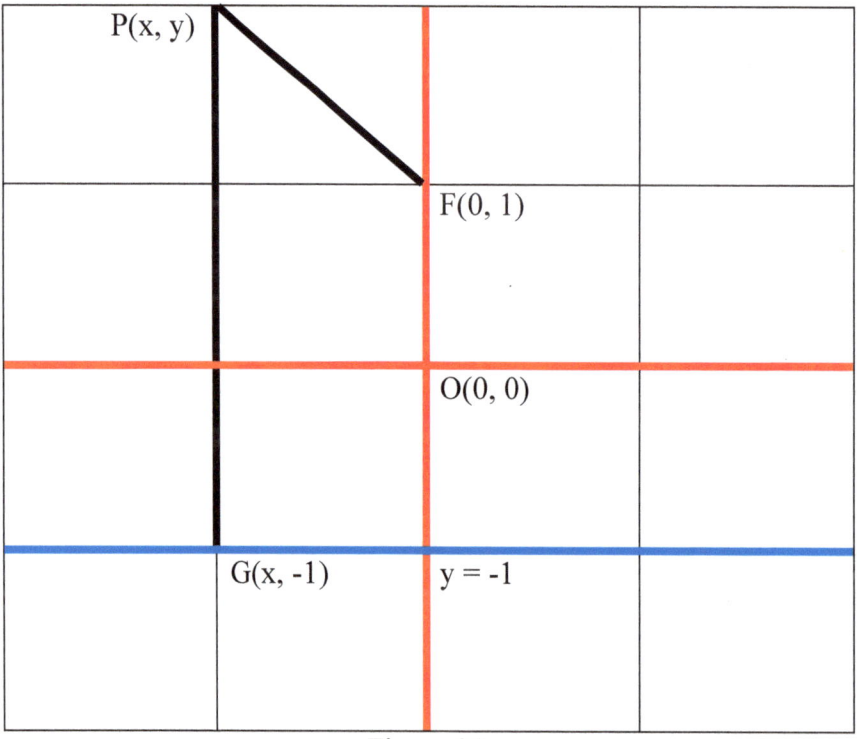

Figure 30

Mentor: Excellent. The distance from P to F, or PF, does not look like the distance from P to G, or PG, but we want these to be equal distances for light to travel. Then all the

40

light leaving the star at the same time will come to F at the same time, as if they all came to the mirror, y = -1 at the same time. We call the mirror line the <u>directrix</u>. Can you write the equation expressing the equality of the two distances?

Madeleine: I think so. PF $= \sqrt{(x-0)^2 + (y-1)^2}$ and PG = y – (-1). Setting those two equal we have the equation:

$$\sqrt{(x-0)^2 + (y-1)^2} = y - (-1)$$

Mentor: Can you simplify this equation?
Madeleine: Oh, yes. I will square both sides to get rid of the square root or <u>radical</u>.

$$(x-0)^2 + (y-1)^2 = (y-(-1))^2$$

and expand the squared expressions

$$x^2 + y^2 - 2y + 1 = y^2 + 2y + 1$$

At this point I can subtract both 1s and y^2s, giving us

$$x^2 - 2y = 2y$$

or

$$x^2 = 4y$$

or

$$y = \frac{1}{4}x^2$$

Mentor: What does this equation describe? Can you find points on the graph?
Madeleine: I'm going to pick values for x that are always even, so I can cancel the fraction 1/4. Here is a bunch: A(2, 1), B(4, 4), C(6, 9), D(8, 16).

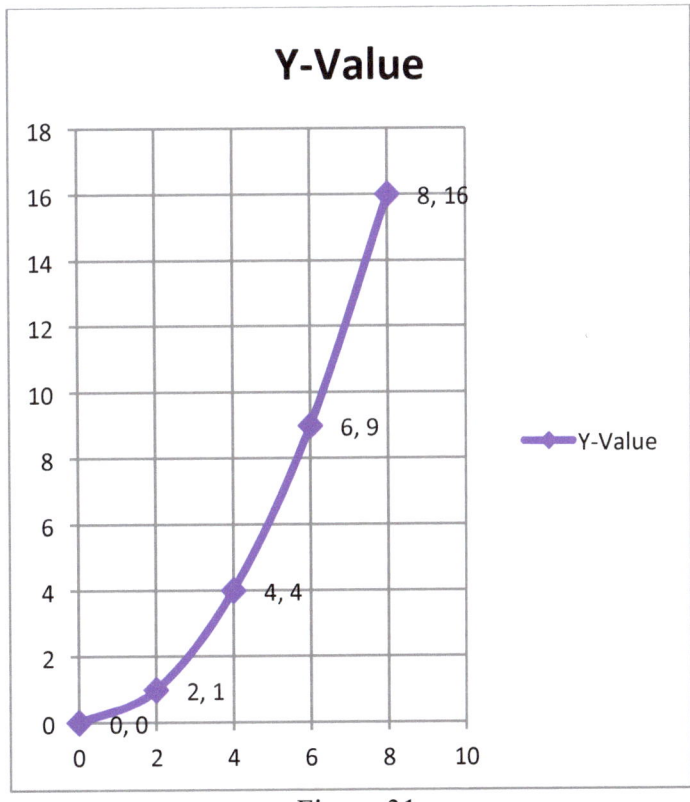

Figure 31

Mentor: Very good, but what about the left side where the x-coordinates are negative?
Madeleine: Oh, yes. I was too busy calculating the values on the right. I should have added these to the list: E(-2, 1), F(-4, 4), G(-6, 9), H(-8, 16).

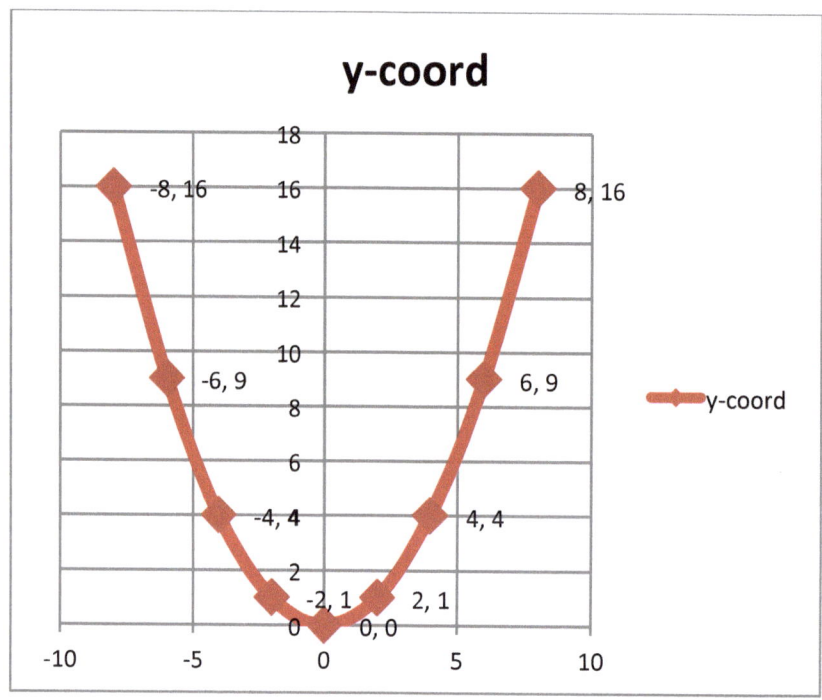

Figure 32

Mentor: So we now see that we can define a <u>parabola</u>, a new geometric figure, in terms of equal distances of any point on the parabola to the focus and to the directrix.

Between the Scenes

1. If we move the focus and directrix closer, how does this change the shape of the parabola? Given F(0, ½) and directrix: y = -1/2.
2. If we move the focus and directrix farther apart, how does this change the shape of the parabola? Given F(0, 5) and directrix: y = -5.

Act 3   Translation Trouble

Scene 1   Motion Problems

Mentor: One of our objectives in this Act centers on transcribing motion described in words into equations and on to numbers. For instance, if I drive my car at thirty miles an hour for two hours, how far will I travel?

Noel: You would travel 60 miles.

Mentor: How do you calculate this number?

Noel: I multiply the number of hours times the rate of travel during that time.

Mentor: How would you represent this action?

Noel: I see a line, really a <u>segment</u>, with one end marked start, "S", and the other end marked finish, "F", as in the figure below.

Figure 33

Mentor: How could we make some of the numbers appear meaningfully along the path?

Noel: We could write the information above the graph.

Speed: 30mph
Time: 2hrs
Distance: 60mi

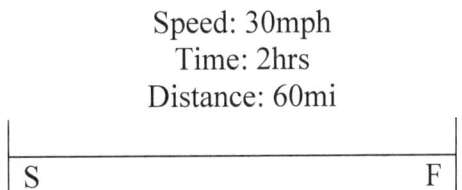

Figure 34

Mentor: We will call the "speed" the "rate." If we use the letters "D", "R", and "T" for the distance, rate and time respectively, could you think of a way to express the relationship between the three quantities?

Noel: Yes. $D = R \times T$.

Mentor: Excellent. Suppose we knew the distance and the time. How would you find the rate?

Noel: We could rewrite the equality as: $R = \frac{D}{T}$.

Mentor: Finally, suppose we knew the distance and the rate, how would you calculate the time?

Noel: $T = \frac{D}{R}$.

Mentor: How are you going to remember all these equations?

Noel: I can remember the first equation with its picture and then change the form by dividing by the appropriate letter.

Mentor: Suppose we are given that a woman drives 100 miles in two hours, what is her rate on the average?

Gus: I divide the 100 miles by 2 hours and get 50 miles per hour.

Mentor: So, if I told you she drove 120 miles at 40 miles per hour, could you tell me how long she was under way?

Gus: Yes. I just divide the distance by the rate to get 3 hours.

Mentor: Exactly. If we can remember that the distance equals the rate times the time, we can arrive at all the other formulas.

Between the Scenes

1. Rob drove to work, a distance of 35 miles, at an average rate of 50 mph. How long did his trip take him?

2. Katty bicycles to the market, a distance of 6 miles, at a rate of 4 mph. How long did it take her to reach the market?

3. Clara left home at 11am headed east; at the same time, Bill started with his car from 450 miles away headed west; both planned to meet somewhere in the middle. If Clara drove at 70 mph; Bill, at 50 mph, how long did they have to drive until they met and how far did each drive?

4. A pilot flies from village A to village B and back in 13 hours with an indicated airspeed of 70 mph. However, flying from A to B, his plane flew against a 20 mph headwind, and flying from B to A he experienced a 10 mph tailwind. How far is village B from village A?

Act 3   Scene 2   Translating Transportation Talk

Mentor: Here we have a typical "word problem" in need of solution. How should we approach this kind of problem? (He hands out problems on a sheet of paper.)

Word Problem 1 (Henry reads aloud.)
Two trains start at the same time from a station in St. Louis, one headed east, the other headed west. The eastbound train travels at 50 mph; the westbound train, at 60 mph. How long will they travel to be 440 miles apart?

Henry: First I will draw a picture of the action. St. Louis is in the middle and the directions are indicated on each side.

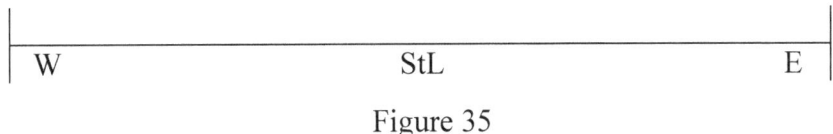

Figure 35

Then I put the information for each train under the picture, <u>letting t equal</u> to the unknown time they both travel.

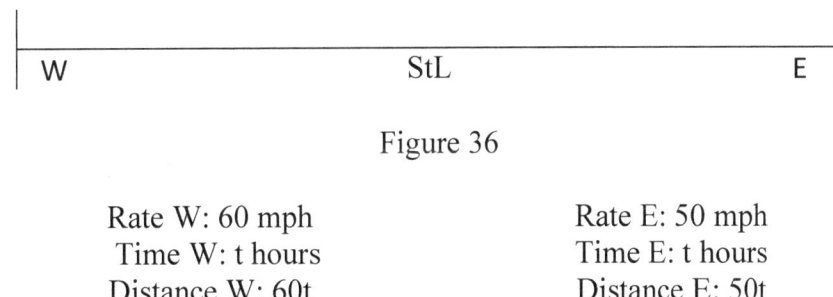

Figure 36

Rate W: 60 mph	Rate E: 50 mph
Time W: t hours	Time E: t hours
Distance W: 60t	Distance E: 50t

Mentor: Excellent! How far apart will the trains be after one hour?
Henry: After one hour the westbound train will have traveled 60 miles and the eastbound train, 50 miles. So combined they will have traveled 110 miles.
Mentor: Can you tell me what the distance between the trains will be after "t" hours?
Henry: The distance between the trains after t hours will be: $d = 60t + 50t$, or $d = 110t$.
Mentor: How long must the trains travel to be 440 miles apart?
Henry: We must solve the equation: $440 = 110t$. Clearly $t = 4$hrs, as $440 = 110 \times 4$.
Mentor: I would like to suggest that we take Polya's (<u>How To Solve It</u>) pattern for solving such problems. He would have us ask three questions:
  1) What is the unknown?
  2) What is the data?
  3) What is the condition?
With these three questions in mind, we can frequently set up an equation and solve it to arrive at a solution.

Word Problem 2 (Emma reads aloud.)
Walking at three miles per hour, Frank leaves home. Two hours later, his sister, Mary, starts after him on her bike, traveling at eight miles per hour. How far will Mary have to bicycle to catch up to Frank?

First Solution:
Emma: I draw a figure, 37, first and then add the data, so I can see what is going on.

Frank's walk before Mary starts   Frank's walk while Mary bikes

Mary bikes

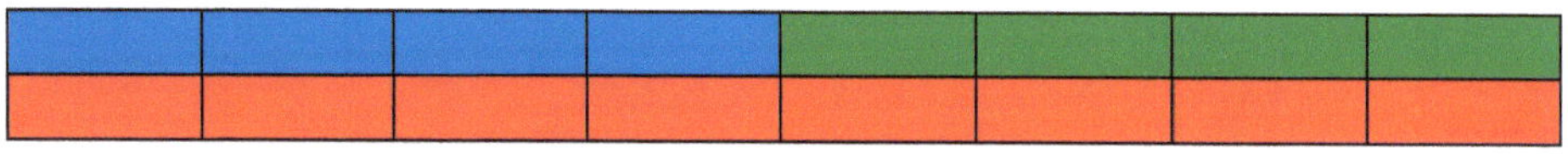

Figure 37

Frank's rate = 3 mph   Let d = miles Frank walks as
Frank walks = 6 miles   Mary bikes.
Frank walks for 2 hours   Then d/3 = hours Frank & Mary go

Frank
Mary

Also d + 6 miles = distance Mary bikes

And $\frac{d+6}{8}$ = number of hours Mary bikes at 8 mph

This gives us the equation of equal times when both are going.

$$\frac{d}{3} = \frac{d+6}{8}$$

Multiplying by 8,

$$\frac{8d}{3} = d + 6$$

Multiplying by 3,

$$8d = 3d + 18$$

Subtracting 3d,

$$5d = 18$$

Dividing by 5,

$$d = \frac{18}{5}, \text{ or } 3\frac{3}{5} \text{ miles}$$

Now add Frank's 6 miles head start,

Mary travels $9\frac{3}{5}$ miles.

Mentor: Excellent solution process. Can anybody see another way to approach this problem?

Second Solution:
Beatrix: I thought of the distance Mary rides as d.
Mentor: Fine. Show us how you use that assumption.
Beatrix: The distance that Frank walks would be 6 miles less than Mary rides once she starts.
Mentor: Please list the parts of this development.

Beatrix: Let d = the distance Mary bikes.

Then $\frac{d}{8}$ = the time Mary bikes

And d – 6 = distance Frank walks while Mary bikes

And $\frac{d-6}{3}$ = time Frank walks with Mary underway

So, we have the equation: $\frac{d}{8} = \frac{d-6}{3}$

Multiplying by 8 and 3, we get

$$3d = 8d - 48$$

Adding 48 and subtracting 3d,

$$48 = 5d, \text{ or}$$

$$d = \frac{48}{5}, \text{ or } 9\frac{3}{5}$$

Mentor: Does anybody else have a different solution?

Third Solution:

Gus: I used the letter 't' for time.

Let t = time in hours Mary bikes to catch up

Then 8t = distance Mary travels in this time

And 6 + 3t = distance Frank travels from start to finish

This gives us the equation

$$8t = 6 + 3t$$

Subtracting 3t,

$$5t = 6, \text{ or}$$

$$t = \frac{6}{5}$$

Multiplying this time by 8 mph,

$$8 \times \frac{6}{5} = \frac{48}{5}, \text{ or } 9\frac{3}{5} \text{ miles}$$

Mentor: Anybody else?

Fourth Solution:

Dave: I let t be the time Frank walked from start to finish. So,

Let t = hours Frank walks

Then 3t = miles Frank walks

And t – 2 = the time Mary rides

And 8(t – 2) = the distance Mary bikes

This gives us the equation

$$3t = 8(t - 2), \text{ or}$$

$$3t = 8t - 16$$

Simplifying $16 = 5t$

So, we have t = 16/5 and 3t = 48/5 or $9\frac{3}{5}$.

Mentor: We have now seen four separate solutions, two using distance as the unknown and two using time as the unknown. We see that we get different solutions to the different equations and might have to do some further calculations to get the answer asked for in the problem.

Between the Scenes:

1. A schoolyard is 50 feet longer than it is wide. If the perimeter equals 220 feet, what are its dimensions?
2. Four years ago a mother was 13 times as old as her son. If she is now 5 times as old as her son, how old is each now?
3. If the perimeter of a rectangular shop is 100 feet and the length is 2 feet longer than the width, what are the dimensions of the shop?
4. Tina leaves the restaurant on her bicycle 6 hours before Mary leaves in her car. If Tina averages 5 mph and Mary averages 32 mph, how long will Mary have to drive to catch up to Tina? How far will Tina have to bicycle?

Act 3   Scene 3   Unknown Rates

Mentor: This time we consider a problem where no rates are given, so we have to set up the equations a little differently.

Word Problem 3 (Madeleine reads aloud.)
Harry and Sue have agreed to deliver their gift to Jane in four hours. If Harry walks the first seven miles to Sue's house at twice the rate that Sue walks to carry the gift the last three miles to Jane's house, how fast does each walk?

First Solution
Madeleine: My first response to this problem is to draw a visual diagram.

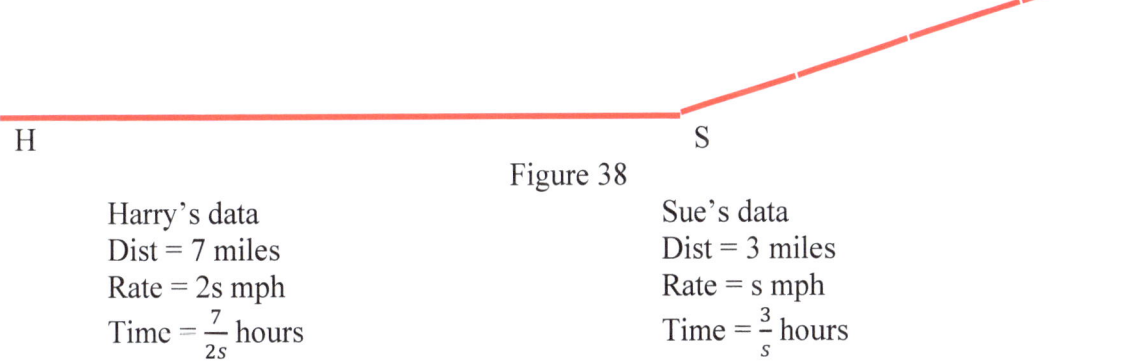

H   S

Figure 38

Harry's data   Sue's data
Dist = 7 miles   Dist = 3 miles
Rate = 2s mph   Rate = s mph
Time = $\frac{7}{2s}$ hours   Time = $\frac{3}{s}$ hours

The horizontal red line shows Harry's path to Sue's house, a seven-mile trip. Sue takes the gift at this point, S, and proceeds to Jane's house, at J. Below the figure, I have entered the data, including Sue's rate of 's' miles per hour. We are told the total time to get the gift to Jane's house is 4 hours, giving us the basis for an equation.

$$\frac{7}{2s} + \frac{3}{s} = 4$$

Multiplying both sides by 2s,

$$7 + 6 = 8s$$

Adding 6 + 7 and dividing by 8,

$$s = \frac{13}{8}$$

We conclude that Sue's rate of travel equals $1\frac{5}{8}$ mph, giving us Harry's rate $2s = \frac{13}{4}$ mph.

Mentor: Madeleine, you have spelled out the process clearly. Did some one solve the problem in a different manner?

Second Solution

Matthew: I drew the same figure, but I made Harry's rate s and Sue's $\frac{1}{2}s$. So I come up with the equation, $\quad \frac{7}{s} + \frac{3}{\frac{1}{2}s} = 4$

Simplifying the second fraction,

$$\frac{7}{s} + \frac{6}{s} = 4$$

Adding the fractions and multiplying by s,

$$13 = 4s$$

So, we have Harry's rate,

$$s = \frac{13}{4}, \text{ or } 3\frac{1}{4} \text{ mph}$$

This is the same result as before. Dividing by 2, we have the same rate for Sue as before.

Third Solution

Noel: I did it differently, using t for Harry's time under way. Here is the data.

Harry	Sue
Let t = time travelling	Then 4 – t = Sue's time travelling
Then $\frac{7}{t}$ = rate	And $\frac{3}{4-t}$ = Sue's rate

Because we are told that Harry walks twice as fast as Sue, we can now create the equation we need in terms of time.

$$\frac{7}{t} = 2 \times \frac{3}{4-t}$$

If we multiply by t and (4 – t), we get

$$7(4 - t) = 6t, \text{ or}$$
$$28 - 7t = 6t,$$

and adding 7t, gives us

$$28 = 13t, \text{ or}$$
$$t = \frac{28}{13}$$

Now we can divide 7 by t, or $\frac{28}{13}$, to get Harry's rate,

$$\frac{7}{\frac{28}{13}} = \frac{13}{28} \times 7, \text{ or}$$
$$\frac{13}{4} \text{ mph}$$

as before!

Mentor: You know how to set up a good solution process! Is there a fourth way?

Georgia: Yes. Just make Sue's time t and Harry's 4 – t. The results will be the same.

Between the Scenes

Mary flies from her local airport 400 miles to the one storing her seaplane that she will use to reach a lake 90 miles farther away. If her first aircraft flies three times as fast as the seaplane and the total flying time is $3\frac{1}{3}$ hours, how fast does each plane fly?

Act 3   Scene 4   Average Rate of Travel Problems

Example: If George travels the first two miles of a trip at 2 mph and the last four miles at a rate of 8 mph, what is his average speed for the whole trip?

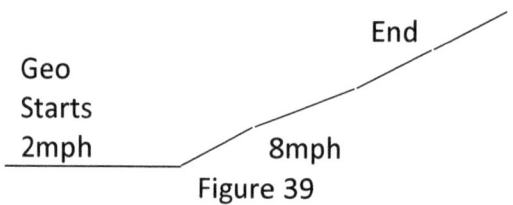

Geo
Starts
2mph   8mph

Figure 39

Solution: Average Rate = Total distance / Total time.

Distance = 2 miles		Distance = 4 miles			
Rate = 2	mph	Rate = 8	mph		
Time = D/R	1 hour	Time = D/R	½ hour		

$$AveRate = \frac{6mi}{\dfrac{2mi}{2mph} + \dfrac{4mi}{8mph}}$$

$$= \frac{6mi}{(1 + \dfrac{1}{2})hr}$$

$$= \frac{6mi}{\dfrac{3}{2}hr}$$

$$= 4mph$$

1.   Abbie drove the five miles from school to the market at 30 mph and then biked the two miles from the market to her house at 10 mph. What was her average rate of travel from school to home?

2.   Betty walked the mile from her house to the bus stop at 3 mph and hopped the bus to work. If the bus covered the two-mile trip at a rate of 15 mph, what was the average rate of Betty's trip?

3.   If Cammie biked the three miles to her brother's at 8 mph and rode the last seven miles to work in her brother's car, how fast would her brother have to drive to make her average rate of travel 15 mph?

4.   If Danny travels the first three miles of a trip at a certain rate and the last five miles at twice that rate, at what rates must he travel to average 10 mph?

52

5.   If Elizabeth travels the two miles from her house to her sister's on her skateboard and rides the remaining five miles at a rate 2 mph faster to school on her sister's motor scooter, how fast does she have to go on each conveyance to average 7 mph for the whole trip?

6.   Following a school bus, Fred averages a certain speed for the first four miles of his way to work. After he gets on the highway, he maintains a rate 20 mph faster for the last 11 miles and has had an average rate for the entire trip of 25 mph. How fast did he travel on each segment of the trip?

Act 4   Quadratic Expressions: Reciprocals & Negative Factors
Scene 1   Connecting the Dots

Mentor: We have seen how the equation of a line, perhaps $y = 3x + 5$, connects all the points in a given list such as $\{(0,5), (1,8), (2,11), (-1,2)\}$. Suppose we are given a non-linear list such as $\{(-5, 25), (-4,16), (-3,9), (-2,4), (-1,1), (0,0), (1,1), (2,4), (3,9), (4,16), (5,25)\}$, where every <u>value</u>, or y-coordinate, is the square of its <u>argument</u>, the x-coordinate. What will this set of points look like?

Matthew: Here is a graph of all those points!

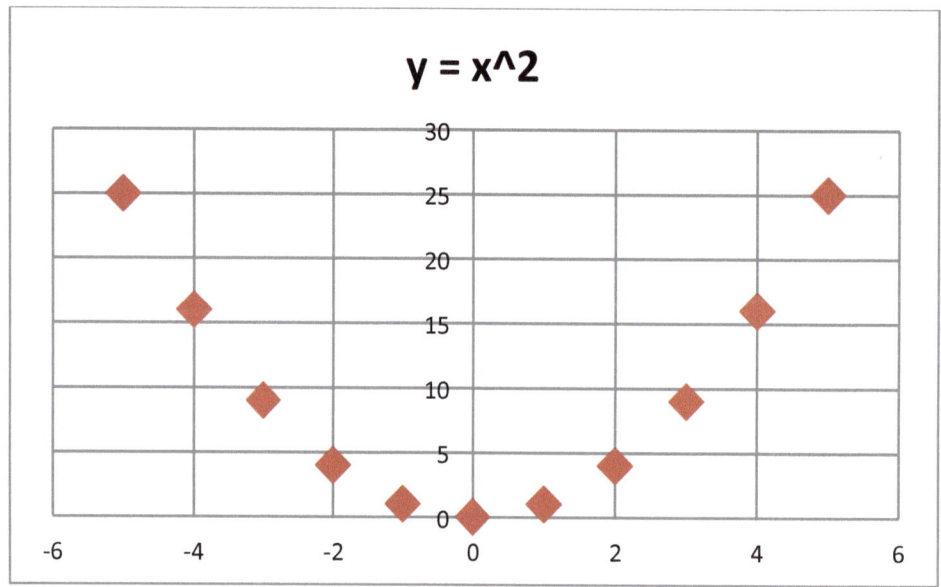

Figure 40

Mentor: How do you think we should join all these points?
Matthew: It looks like a big smile!

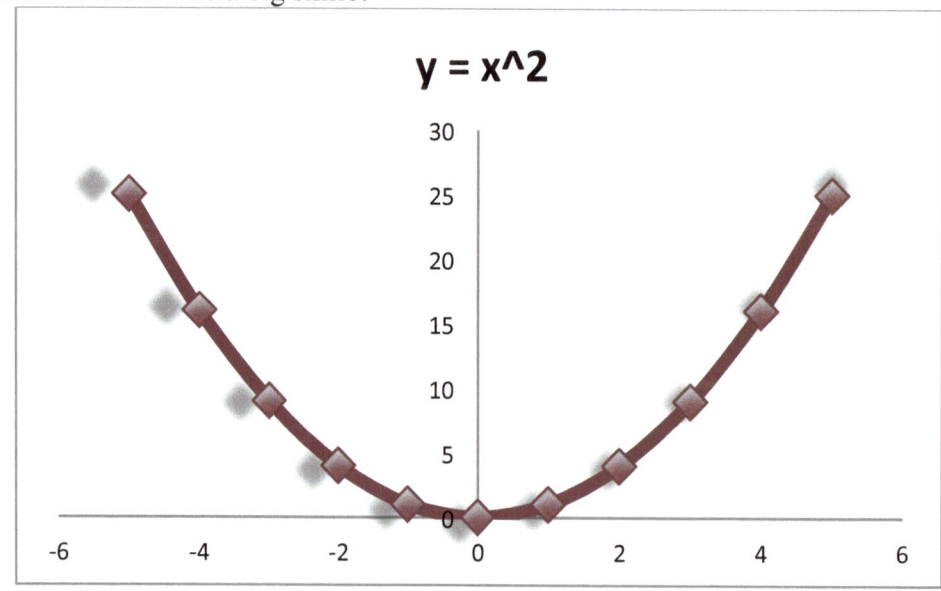

Figure 41

54

Mentor: Now suppose we add 4 to each square, so that we have changed the points to (-5, 29), and (0, 4) and (5, 29). What has happened to your smile?

Matthew: All the points are lifted 4 units higher on the coordinate system. It looks like this.

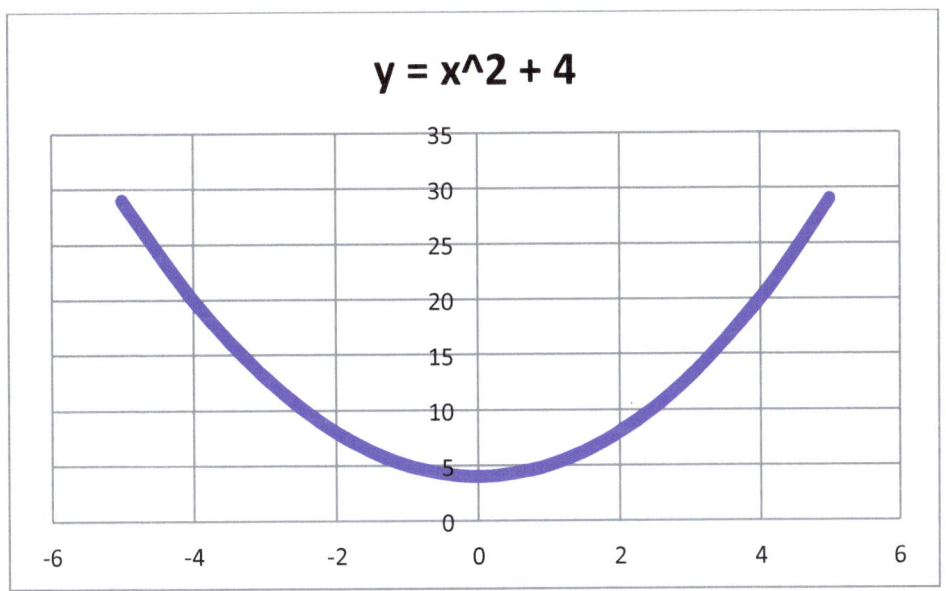

Figure 42

Mentor: Where does this curve cross the x-axis?

Matthew: It doesn't, because the lowest value equals 0 + 4.

Mentor: Now show me what will happen, if I subtract 4 instead of adding 4.

Matthew: Easy. Just slide the original smile down 4 units. Here!

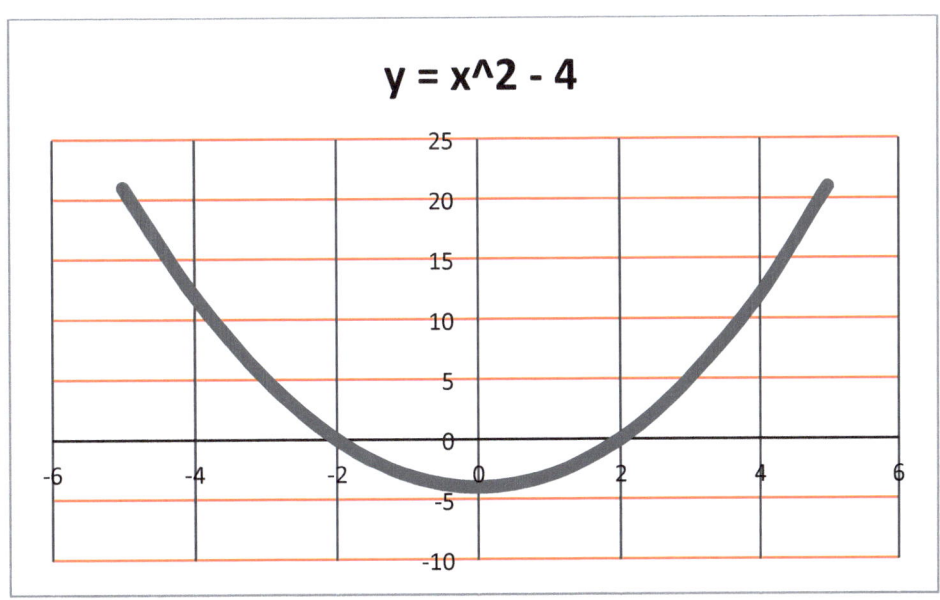

Figure 43

Mentor: Excellent! Where does this curve cross the x-axis?

Matthew: I can read this graph. At -2 and 2!

55

Mentor: So what kind of an expression do we have in this quadratic form?

Matthew: Oh, I see. We have the difference of two squares: x^2 and 4!

Mentor: What can we do with expressions like this?

Matthew: We can factor the difference of two squares: $(x-2)(x+2)$. The first factor equals zero when $x = 2$ and the second equals zero when $x = -2$.

Mentor: Good. Now you can see how learning to factor the difference of two squares in our arithmetic work makes drawing quadratic expressions much easier. We call these "smiles" <u>parabolas</u>.

Between the Scenes

Draw the parabolas below. Where does each of these parabolas cross the x-axis, if it does at all?

1) $y = x^2 + 1$
2) $y = x^2 - 1$
3) $y = x^2 + 9$
4) $y = x^2 - 9$
5) $y = x^2 - 3$

Act 4   Scene 2   Multiplying by a Constant

Mentor: What happened to the graph of a line when we multiplied by a large slope, say 6 in a line y = x + 7, so y = 6x + 7?

Emma: That's easy after Act 1 Scene 5. We see that the line appears to "stretch" up on the positive values and "stretch" down on the negative values. If we multiplied the slope by a negative number, say -6, the line appears to stretch up on the left of the x-intercept and down on the right side of the x-intercept.

Mentor: Using that as background, what will happen to the graph of y = x^2 if we multiply by 5? So we have y = $5x^2$.

Emma: All the positive points will move 5 times as high as the corresponding points on the original curve.

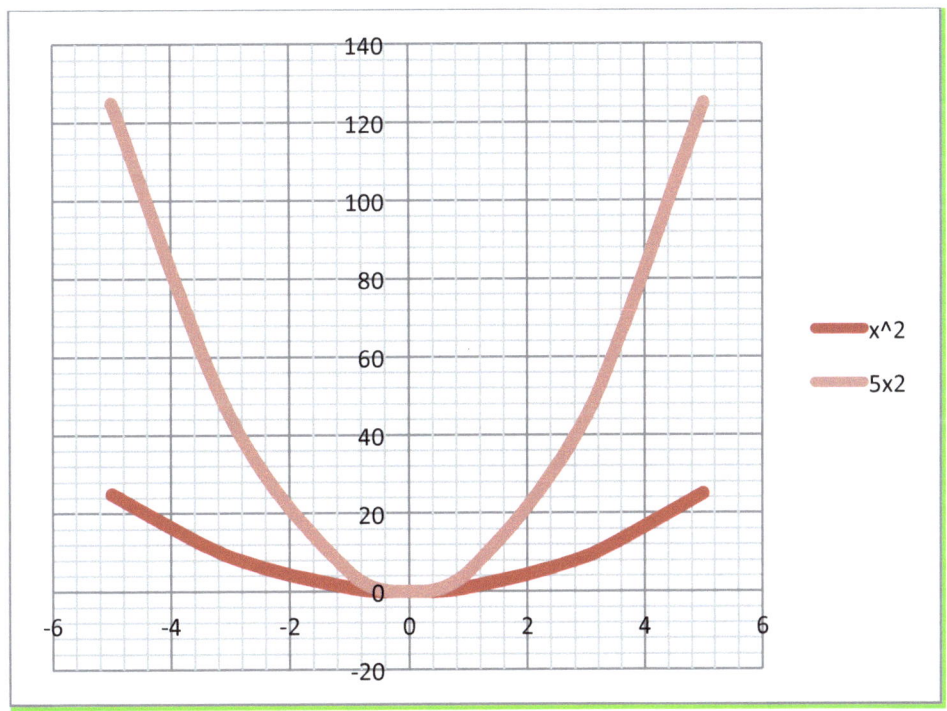

Figure 44

Mentor: You have got the picture! Now tell me what would happen, if we multiplied by -5 instead of by 5.

Emma: I would just turn the picture upside down as in Figure 45! You can see that the light pink curve, Series 3, is the mirror image of the top parabola, Series 2.

Mentor: What rule would you follow to describe the graph of the curve with the sign changed from plus to minus?

Emma: I would see the new curve as the mirror image about the x-axis of the old curve. What we see below the x-axis will appear above it, and what we see above the x-axis will appear below it.

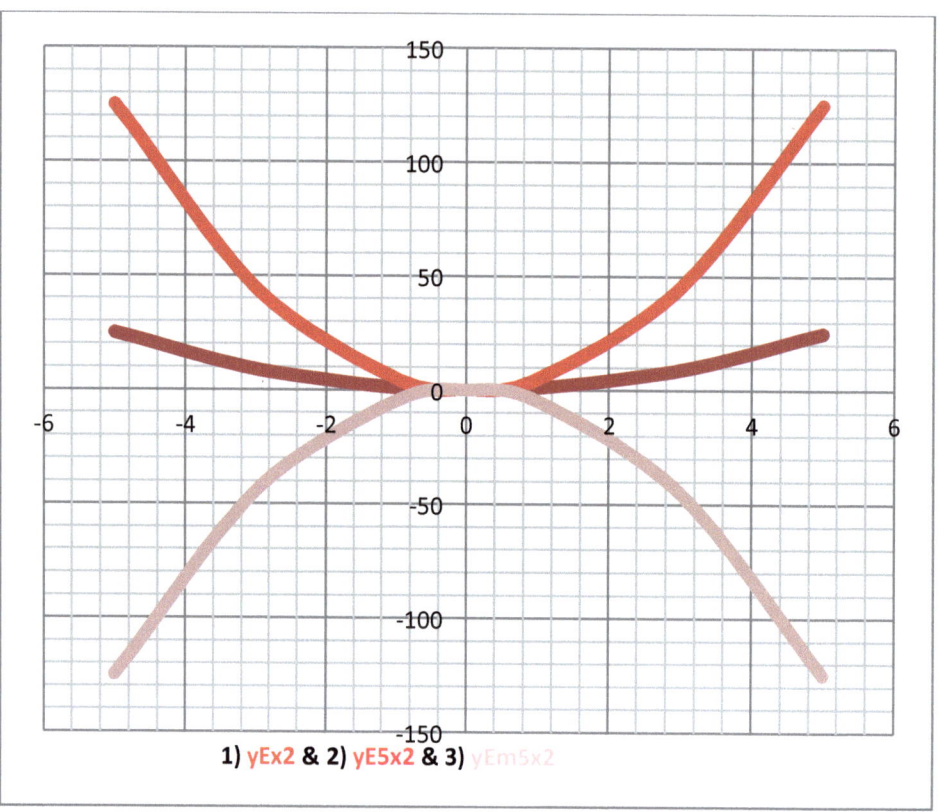

1) yEx2 & 2) yE5x2 & 3) yEm5x2

Figure 45

Between the Scenes

Draw the parabolas below. Where does each of these parabolas cross the x-axis, if it does at all? Do you see any common factors?

1) $y = -7x^2 - 7$
2) $y = 7x^2 - 7$
3) $y = -3x^2 + 3$
4) $y = 3x^2 + 3$
5) $y = -5x^2 - 45$
6) $y = 5x^2 - 45$
7) $y = -5x^2 + 45$
8) $y = 5x^2 + 45$
9) $y = -4x^2 + 12$
10) $y = 4x^2 - 12$

Act 4   Scene 3   Dividing by x

Mentor: Up to now we have avoided dividing by x, but we shall discover a new shape and its application to a number of different expressions. We go back to our first graph of the line y = x, looking at the x-coordinates and their reciprocals. Your assignment here is to find the points on $y = \frac{1}{x}$ corresponding to points on y = x.

Georgia: Here is my graph of the two expressions.

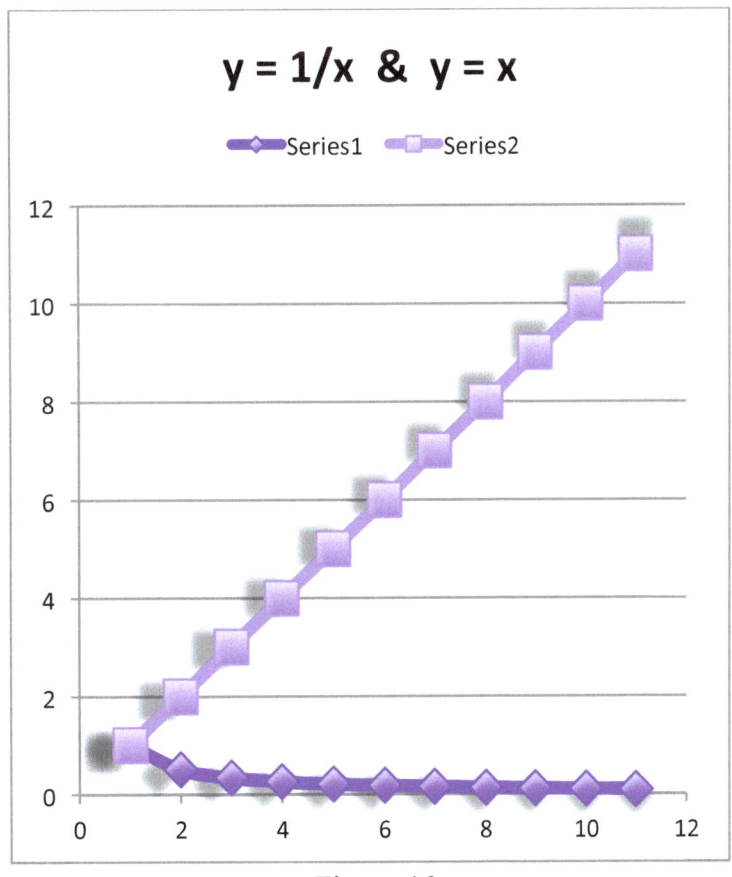

Figure 46

Mentor: Excellent. Can you describe the two graphs and their relationship to each other?

Georgia: Yes. We know that y = x starts at the origin and makes a line up the diagonal between the two positive axes. And what I see is that the reciprocal points start at (1,1) and get closer and closer to the x-axis, going down!

Mentor: That's good along the positive x-axis for large arguments for x, but what happens to the graph for x between 0 and 1, or 0 < x < 1?

Georgia: Just as the reciprocal of a large number will be a small number, the reciprocal of a small number will be a large number. Here is a graph with the lighter color, the y = x line, and the darker curve, its reciprocal. See Figure 47.

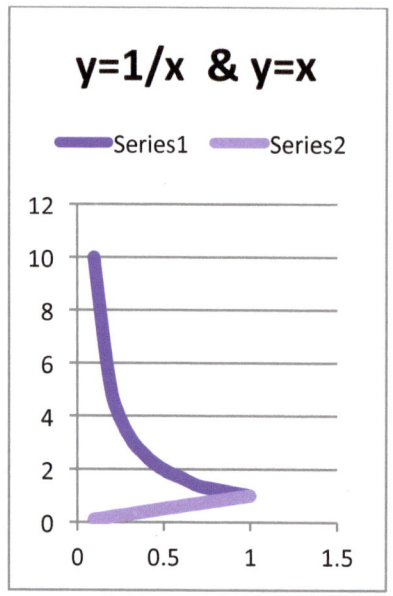

Figure 47

Mentor: Looking at the graph we have drawn in Figure 47, what do you think should be drawn in the third quadrant?

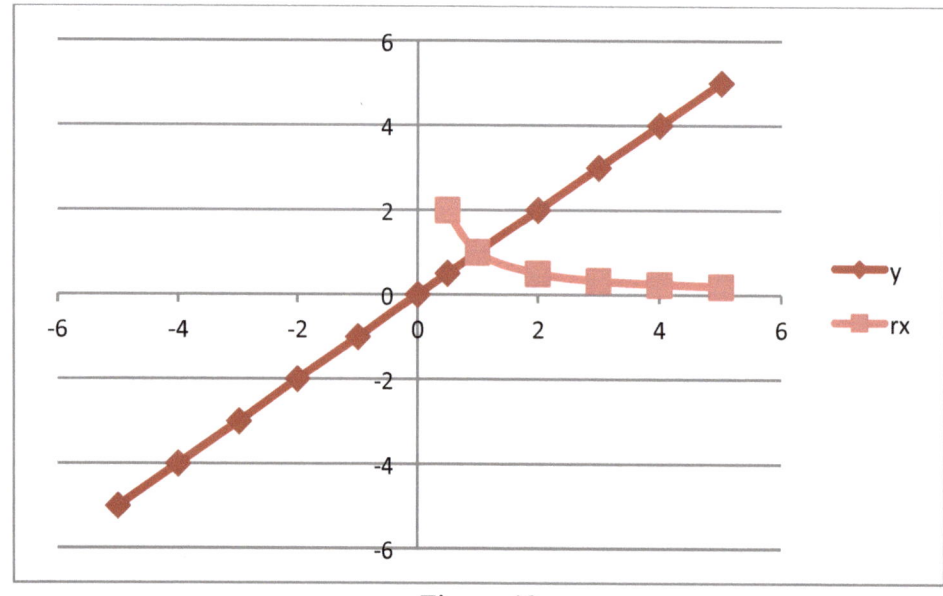

Figure 48

Dmitri: The third quadrant should look like the first quadrant seen through a lens at the origin. Out on the left, the reciprocal will be close to the negative x-axis and go down as the x-coordinates are taken closer to the origin. See Figure 49.

60

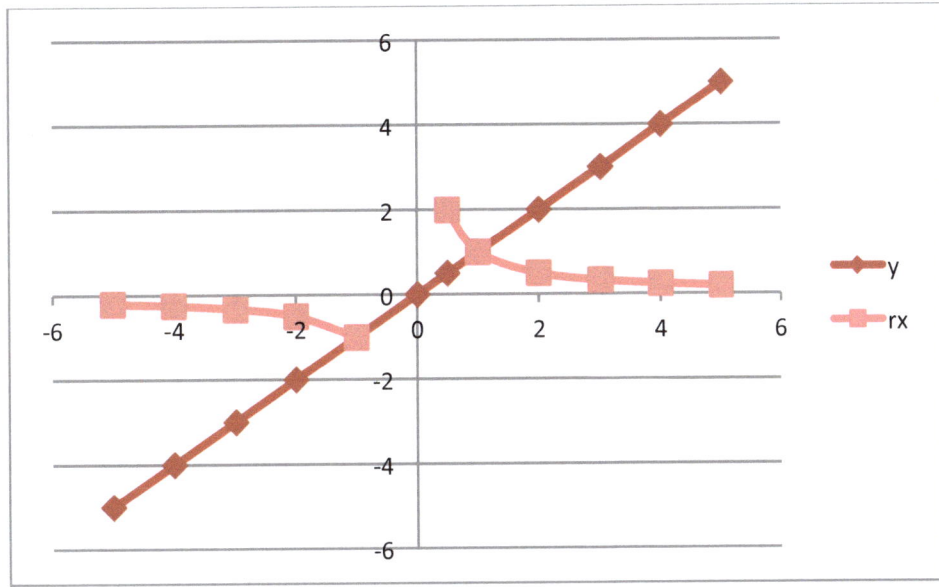

Figure 49

Mentor: Excellent description and graph. In the next scene we will look at a more complicated, but very similar, curve. To suggest the way <u>asymptotes</u> might look, we show $y = \frac{x}{x^2-1}$ with the almost vertical lines at x = -1 and x = 1 in Figure 50.

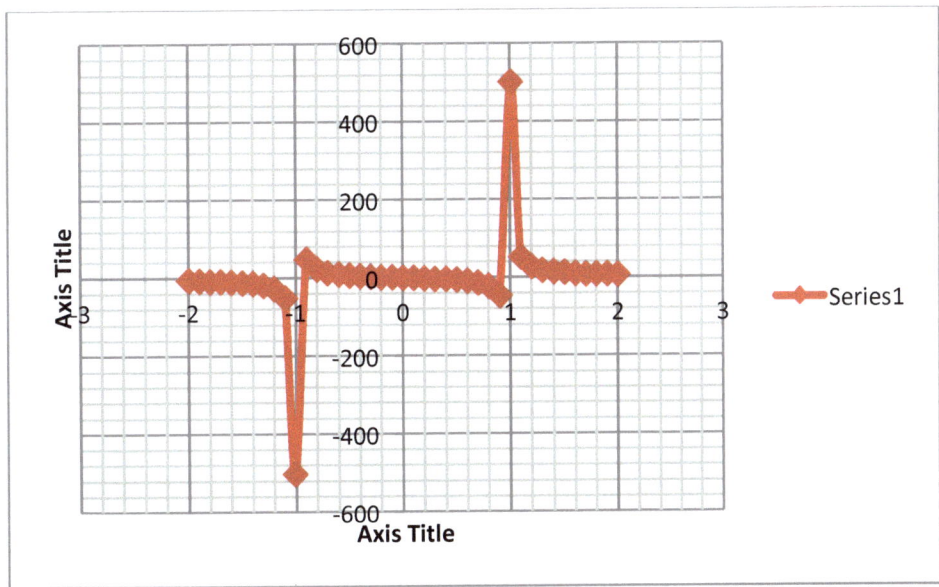

Figure 50

Between the Scenes
Sketch the graph of each of the following curves and describe the changes.
1.   y = x,   y = -x and   y = -1/x.
2.   y = x,   y = -3x and y = -1/(3x).
3.   y = x,   y = x – 2 and $y = \frac{1}{x-2}$.
4.   y = x,   y = 3x – 6 and $y = \frac{1}{3x-6}$.

61

Act 4   Scene 4   Dividing by x² – 9

Mentor: We have begun to see how the reciprocal of a curve relates to the given curve, where the given curve is a line. In this scene we will examine the curves related to a quadratic expression: $y = x^2 - 9$. What do we already know about such a curve?

Dmitri: If we factor the difference of two squares, we will see the two x-intercepts, where we cannot calculate reciprocals. So we graph $y = (x - 3)(x + 3)$ in Figure 51.

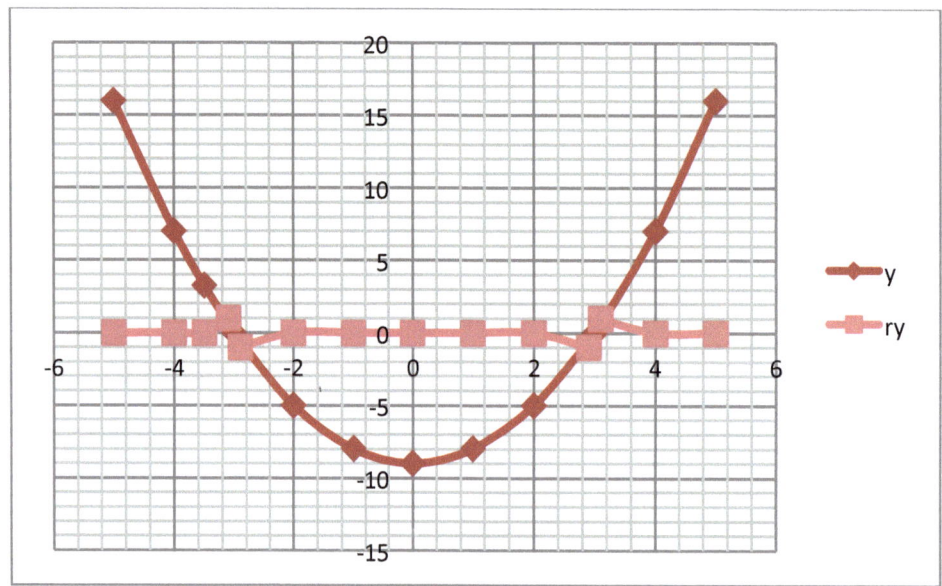

Figure 51

Mentor: What can we see from your graph?

Dmitri: Approaching -3 from the left side, we have $y = x^2 - 9$ (indicated by x2m9) getting closer to the x-axis from above and the 1/y (indicated by recy) starting to turn up and going away from the x-axis.

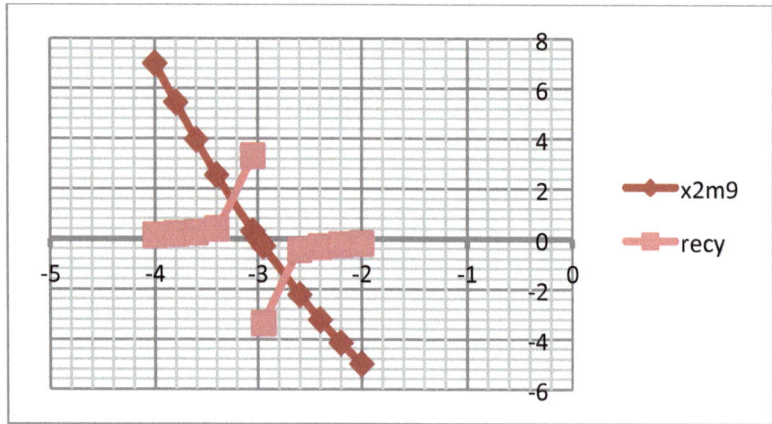

Figure 52

62

Mentor: If we had a vertical line x = -3, called an <u>asymptote</u>, we would find that the <u>reciprocal would get closer to its asymptote the higher we graphed it on the left and the lower we graphed it on the right</u> of x = -3.

Let's consider the part of the parabola below the x-axis, between x = -3 and x = 3. What will its reciprocal look like?

Mat: I think it will look like the lower part of the parabola turned upside down.

Mentor: I have a caving metaphor to describe this situation. In a cave, stala**C**tites hang down from the **C**eiling because the water drips through and brings material with it, and the drips on the **g**round build up stala**g**mites. So <u>for every stalactite hanging out of the ceiling there is a stalagmite growing up out of the ground</u>.

In the next scene, we will look at the reciprocal of a parabola that does not touch the x-axis.

Between the Scenes

Sketch each of the curves below and their reciprocals.

1. $y = x^2 - 16$
2. $y = 5x^2 - 20$
3. $y = (x - 2)(x - 6)$
4. $y = (x + 2)(x + 6)$
5. $y = -7(x + 2)(x + 6)$
6. $y = (x + 5)(x - 4)(x - 8)$
7. $y = (x + 3)^2(x - 9)$

Act 4   Scene 5   The Reciprocal of $x^2 + 9$

Mentor: In this scene we can see what a difference it makes for a polynomial curve to intersect or not intersect the x-axis. Here we show $y = x^2 + 9$ in two ways. The first, in Figure 53, is numerically accurate, but makes the reciprocal look like a straight line.

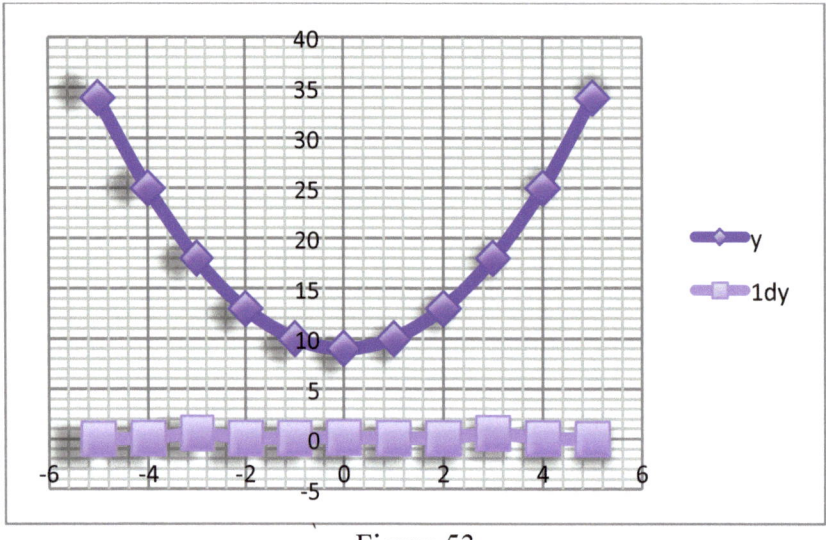

Figure 53

The next picture, Figure 54, shows the reciprocal with all its y-coordinates multiplied by 10. Clearly this is a distortion, but we are made aware of the changes. How could we

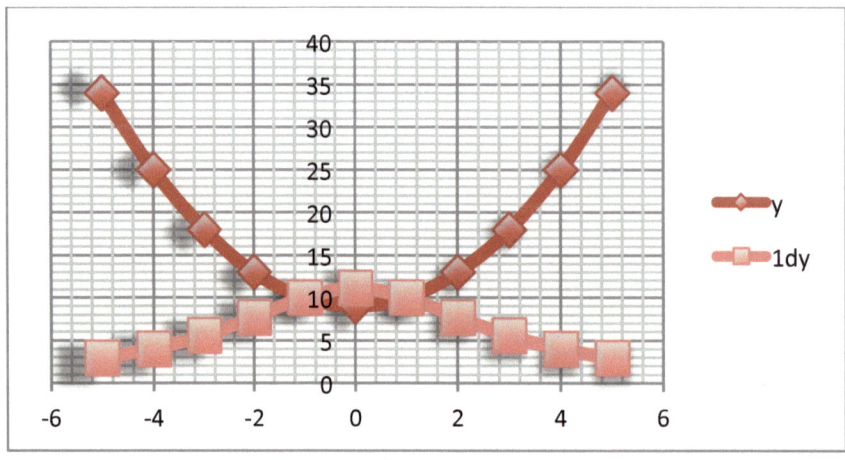

Figure 54

describe the relationship between the original parabola and its reciprocal?

Henry: It looks like the stalactite and stalagmite all over again!

Mentor: How is this picture different from the one of the parabola that intersects the x-axis in two places? (Note: "1dy" means "1 divided by y.")

Henry: The stalagmite appears to level out, to the left and to the right, on top of the x-axis rather than being squeezed between two vertical lines or <u>asymptotes</u>.

Mentor: So what do we do when we draw the reciprocal of such a parabola?

Henry: Distort, or exaggerate, the bump corresponding to the bottom of the parabola.

Mentor: What does the parabola $y = (x - 4)^2 + 5$ and its reciprocal look like?

Henry: The smallest value of $(x - 4)^2$ is zero, when x = 4. So the lowest point on the parabola is at (4, 5).

Mentor: We say that the parabola has been <u>shifted</u> to the right 4 units. The figure does not change shape, just its position has moved along the positive x-axis. What do the graph and its reciprocal look like?

Henry: They look like Figure 54 moved to the right 4 and with the vertex of the parabola at (4,5). Here is my drawing with the distorted reciprocal.

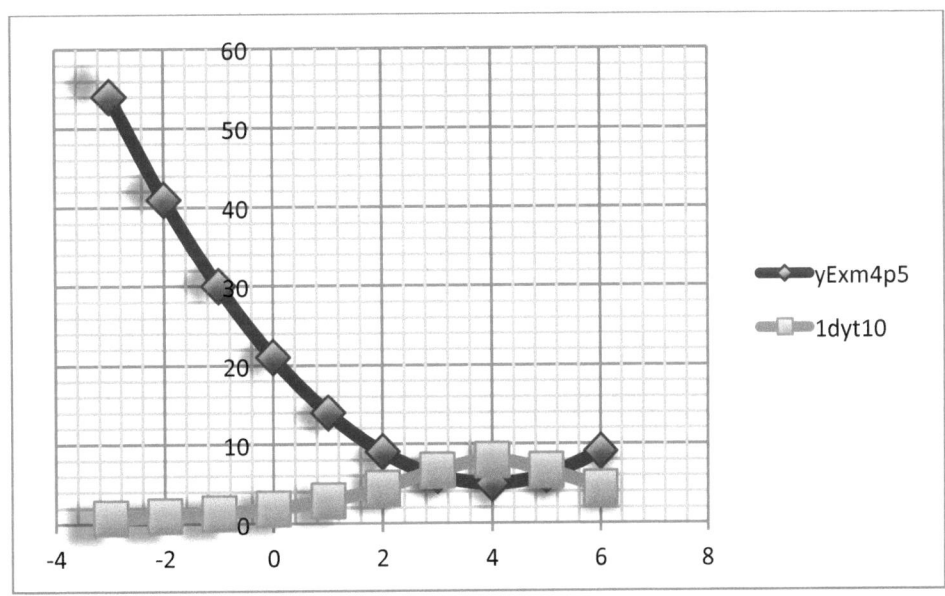

Figure 55

Mentor: When we graphed the two lines 1) y = x and 2) y = -3x + 20, what changes did we make to the first graph to get the second?

Henry: We raised the first line up 20 units and "rotated" the right side down and the left side up.

Mentor: So how do we describe what we do to $y = x^2$ to get a picture of
$$y = (x - 3)^2 + 8?$$

Henry: Slide the first graph to the right 3 and shift it up 8.

Mentor: Excellent.

Between the Scenes

Sketch and describe the curves a) $y = (x - 4)^2 + 9$ and b) $y = (x + 4)^2 - 9$. Why do their reciprocals look so different?

Similarly, consider the curves c) $y = 6(x + 5)^2 + 16$ and d) $y = 7(x - 2)^2 - 25$.

Act 5   The Quadratic Equation

Scene 1   The Transition y = 5x - 2 to y = 5(x – 2)

Mentor: Earlier we have seen how rewriting the equation of a line, y = x, to y = x – 3 shifts the line around in the plane. How do we look at this shift?
Georgia: I would draw the two lines first!

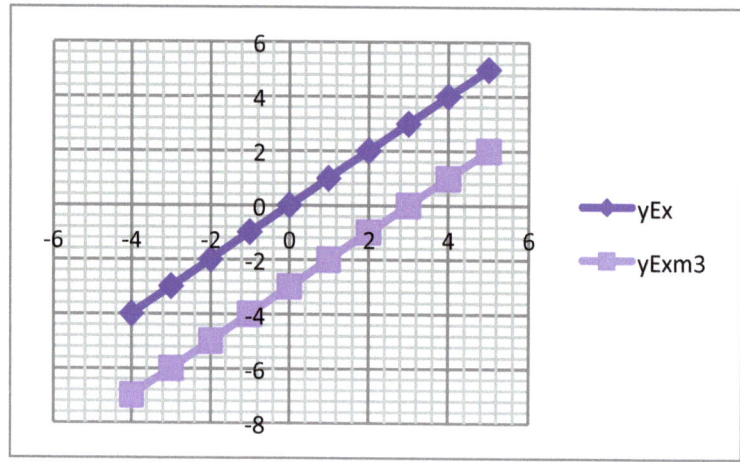

Figure 56

We can see that the y = x, or yEx in purple, graph gets shifted down 3 units to make the graph of y = x – 3, or yExm3 in pale purple.
Mentor: Nice. However, the situation is ambiguous. How else could we see this shift?
Georgia: Oh, yes. We could also see the pale line being shifted to the right 3 units.
Mentor: Good. Let's try to see the distinction by making the formula a little more complicated with a change in slope from 1 to, say, 4. We can write these two curves in slightly different ways, and that should help us see an important difference. How will the graph of y = 4x – 3 differ from the graph of y = 4(x – 3)?
Georgia: Here are the graphs of the two lines in Figure 57 on the next page.
We can see from the graph that the y = 4x – 3 line, in purple, intersects the y-axis just below the x-axis, so at (0, -3). This represents a shift of 3 down on the y-axis from the line of y = 4x, not shown. On the other, pale curve, we see that it intersects the x-axis at (3,0). So we conclude that y = 4x has been shifted 3 units to the right. This pale line intersects the y-axis below y = -10, actually at (0, -12).
Mentor: Good work! The different scales on the x- and y-axes make reading the values harder, but you have seen it correctly. And calculated it also!
Now looking at the two equations, how can we describe the difference?
Georgia: We know the order of operations is: ^, ×, ÷, +, −, unless we use parentheses to require a subtraction before a multiplication as shown in y = 4(x – 3).
Mentor: Excellent! Now we can use some new vocabulary. The first operation we perform we call the <u>inner operation</u>, the last operation we perform we call the <u>outer operation</u>. Can you tell me the inner and outer operations in the two equations?

66

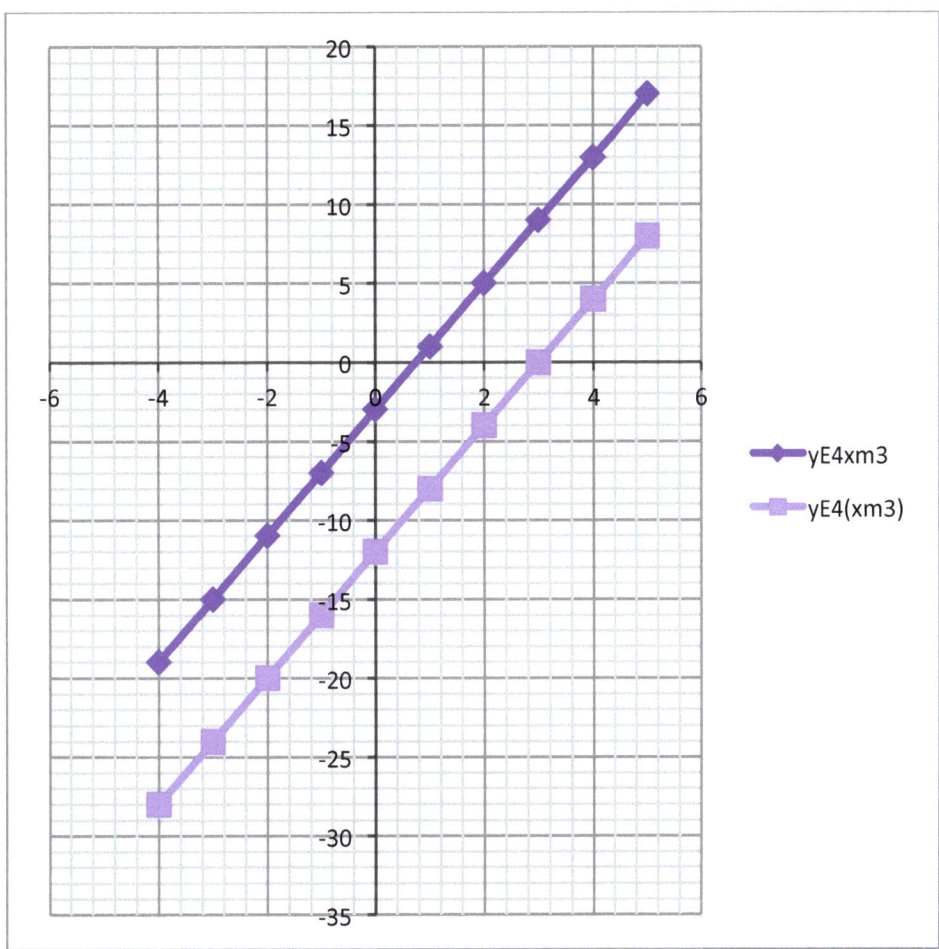

Figure 57

Georgia: In y = 4x – 3, the inner operation is multiplication; the outer, subtraction. In y = 4(x – 3), the inner operation is subtraction; the outer, multiplication. The order has been reversed!

Mentor: Nice. Now tell us what happens to the graph of y = 4x, when we rewrite it as we have in 1) y = 4x – 3 and in 2) y = 4(x – 3)?

Georgia: If the subtraction is the outer operation, as in equation 1), then the curve shifts down 3. If, on the other hand, the subtraction comes as the inner operation, the curve shifts across 3 to the right.

Mentor: We have made a big and important jump. It prepares us well for the next set of quadratic curves.

Between the Scenes
Graph the following pairs of equations, clearly indicating the direction of shift.
    1)   y = 3x – 4 and y = 3(x – 4)
    2)   y = 5x + 6 and y = 5(x + 6)
    3)   y = -2x + 7 and y = -2(x + 7)
    4)   y = -6x + 1 and y = -6(x + 1)

Act 5   Scene 2   Shifting Quadratic Functions

Mentor: In Scene 1, we saw the importance of the order of operations to indicate the placement of the curve under consideration. Let us observe how the curve of $y = x^2$ changes by including a subtraction: 1) $y = x^2 - 4$ and 2) $y = (x - 4)^2$. Identify the inner and outer operations as you go.

Emma: In the first, we square x and then subtract 4, making multiplication the inner operation. In the second, we subtract 4 first and then square the expression, making subtraction the inner operation.

Mentor: Fine. Now how does this difference affect the graphs of each?

Emma: When the subtraction is the outer operation, the curve moves down, but when it is the inner operation it moves across to the right. Here are my graphs.

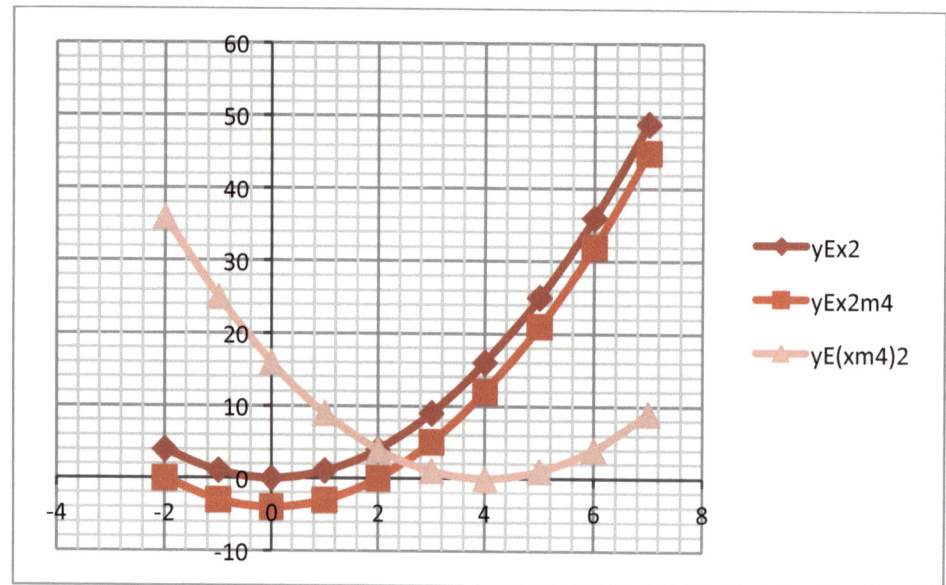

Figure 58

The diamond curve represents the equation $y = x^2$; the square, $y = x^2 - 4$; the pink triangle, $y = (x - 4)^2$.

By subtracting 4 after squaring, making it the outer operation, we move the diamond curve down to the square curve. By subtracting 4 first, now as the inner operation, we shift the diamond curve to the right 4, as shown in pink.

Mentor: Another bit of vocabulary. We call the points on a curve that intersect the x-axis the zeros of the curve. This name fits, because the y value at any of those points is zero. What are the zeros of each of the above curves?

Emma: The zero of $y = x^2$ is 0. The zeros of $y = x^2 - 4$ are -2 and 2. Finally, the zero of $y = (x - 4)^2$ is 4.

Mentor: Well done.

Between the Scenes

Graph and identify the inner and outer operations of each of the following curves.

1) $y = x^2$, $y = 3x^2$,   and $y = (3x)^2$
2) $y = 5x^2$, $y = 5x^2 - 20$ and $y = 5(x - 20)^2$
3) $y = 7x^2$, $y = 7x^2 - 28$ and $y = 7(x - 28)^2$

68

Act 5   Scene 3   The Trinomial Square

Mentor: In Act 2 Scene 2 and in <u>You Can Count On It</u>, we have developed the idea of the difference of two squares. Now we want to look at the "trinomial square," so named because it has three, tri, terms. What does $(x-3)^2$ look like without parentheses?

Maddy: We have learned to graph or picture algebraic expressions to gain some understanding of them. So that is what I will do! I have taken x to be equal to 8 for this figure.

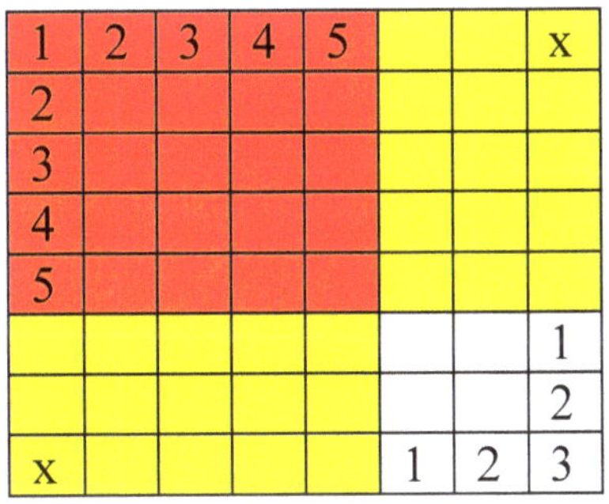

1	2	3	4	5			x
2							
3							
4							
5							
							1
							2
x					1	2	3

Figure 59

In the Figure 59, the red square has $x-3$ cells on each side, so its area is $(x-3)^2$. We see that the whole figure has an area of x^2. So we have to take away the yellow and white area from x^2 to the formula for just the red square. If I take away the yellow column with the white bottom, I have $x^2 - 3x$ left. If I take away the yellow row with the white right corner, I will have $x^2 - 3x - 3x$. We have to note that I have removed the white 3-by-3 square twice! Therefore, I will have to add one white region back in: $x^2 - 3x - 3x + 3^2$. Adding the two -3x's together, I will have $(x-3)^2 = x^2 - 6x + 9$.

Mentor: Whatever happened to the 3?

Maddy: The 3 is hidden in the 6 and the 9.

Mentor: If I gave you $x^2 - 8x + 16$, could you write it as the square of something?

Maddy: Oh, yes. I can see that 4 is hidden in both 8 and 16, $8 = 2 \times 4$, and $16 = 4 \times 4$.

Mentor: So how would you factor $x^2 - 8x + 16$?

Maddy: $x^2 - 8x + 16 = (x-4)^2$

Mentor: Now let's change the sign to $(x+4)^2$. What would the expansion of this product be?

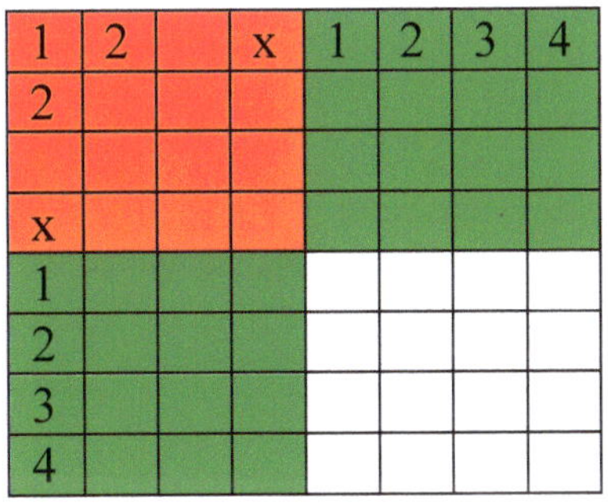

1	2		x	1	2	3	4
2							
x							
1							
2							
3							
4							

Figure 60

69

Maddy: It would be

$$(x + 4)^2 = (x + 4)(x + 4)$$
or distributing the first $(x + 4)$ over sum of the second
$$(x + 4)x + (x + 4)4,$$
distributing again
$$x^2 + 4x + 4x + 16,$$
collecting like terms
$$x^2 + 8x + 16,$$
so
$$(x + 4)^2 = x^2 + 8x + 16.$$

Mentor: How would you describe what you did, using $(x + 7)^2$ as an example?

Maddy: $(x + 7)^2 = x^2 + 14x + 49$ by squaring the x, multiplying the x by 2 times the 7, and finally squaring the 7.

Mentor: Good work! How would you express the square $(2x - 5)^2$ as a sum?

Maddy: Square the 2x, making $4x^2$, multiply 2 times -10x, and square 5. The result would be $4x^2 - 20x + 25$.

Mentor: I think you have the pattern!

Between the Scenes

Express each of the squares below as a trinomial square and graph their curves, indicating the coordinates of the vertex as in Scene 2. You may want to take out a common factor first.

1) $(x - 5)^2$
2) $(x + 5)^2$
3) $(x - 10)^2$
4) $(x + 10)^2$
5) $(2x - 3)^2$
6) $(2x + 3)^2$
7) $(3x - 2)^2$
8) $(3x + 2)^2$
9) $(5x - 15)^2$
10) $(7x - 28)^2$

Act 5   Scene 4   Factoring the Trinomial Square

Mentor: Now that we have learned how to expand the binomial squared, we can try reversing the process by factoring the trinomial square. Factor $x^2 + 10x + 25$.

Ursula: Easy: $x^2 + 10x + 25 = (x + 5)^2$.

Mentor: Tell us what you did.

Ursula: I divided the coefficient of the x term, 10, by 2 to get 5 and checked that the third term was equal to 5^2.

Mentor: Fine. How would you have changed your result if you had been given $x^2 - 10x + 25$?

Ursula: I would have changed my middle term to -10x from + 10x, and then divided -10 by 2, to arrive at -5. Thus we get $x^2 - 10x + 25 = (x - 5)^2$

Mentor: If I wanted to "complete the square" of the expression $x^2 + 6x$, what would the third term be?

Ursula: I would divide 6 by 2 and square the result to get 9. I get $x^2 + 6x + 9$.

Mentor: You make it sound easy! How would you factor $x^2 + 6x + 9$?

Ursula: I say $x^2 + 6x + 9 = (x + 3)^2$.

Mentor: So, if you had to complete the square of $x^2 + bx$, what would you do?

Ursula: I would divide b by 2 and square it, arriving at: $x^2 + 2(\frac{b}{2})x + (\frac{b}{2})^2 = (x + (\frac{b}{2}))^2$.

Between the Scenes

Factor the following Trinomial Squares, sometimes hidden by a common factor in each term. Graph each curve, indicating the location of the vertex.

    1) $x^2 - 12x + 36$
    2) $x^2 + 12x + 36$
    3) $x^2 - 14x + 49$
    4) $x^2 + 14x + 49$
    5) $4x^2 - 24x + 36$
    6) $6x^2 - 24x + 24$
    7) $5x^2 - 70x + 245$
    8) $5x^2 + 70x + 245$
    9) $x^2 + 2bx + b^2$
    10) $x^2 - 2dx + d^2$

Complete the following Trinomial Squares, allowing for the common factors.

    11) $x^2 - 4x + ?$
    12) $x^2 + 4x + ?$
    13) $x^2 - 6x + ?$
    14) $x^2 + 7x + ?$
    15) $x^2 - 3x + ?$
    16) $x^2 + 11x + ?$
    17) $x^2 - 13x + ?$
    18) $3x^2 - 12x + ?$
    19) $5x^2 + 10x + ?$
    20) $7x^2 - 14x + ?$

Act 5   Scene 5   Solving a Quadratic Equation I

Mentor: We have learned how to complete the square and factor the difference of two squares, so let's put the two together to solve a quadratic equation: $x^2 - 6x + 5 = 0$.

Anna: I don't see a perfect square, so I have to add 4 to the five to make $x^2 - 6x + 5$ into the square $x^2 - 6x + 9$ with the compensation of subtracting 4. This gives me the equation: $x^2 - 6x + 9 - 4 = 0$.

Mentor: Now what do you have to do?

Anna: I factor the first three terms so I can see that I have the difference of two squares.

$$(x - 3)^2 - 4 = 0$$
$$\text{or}$$
$$(x - 3)^2 - 2^2 = 0$$

Now I can factor this difference in the usual way we have seen before.

$$\{(x - 3) - 2\}\{(x - 3) + 2\} = 0$$
$$\text{or}$$
$$\{x - 5\}\{x - 1\} = 0$$

This gives us the two equations: $x - 5 = 0$ and $x - 1 = 0$, which we easily solve:

$$x = 5 \text{ or } x = 1$$

Mentor: Let's try another variation following the sequence above.

Solve for x: $x^2 - 10x + 21 = 0$

Caroline: First I look at the first two terms, $x^2 - 10x$ to see what the third term must be to complete the square. Dividing 10 by 2, I realize that the third term has to be $5^2 = 25$. So I have to add 25 and then subtract it back out again.

$$x^2 - 10x + 25 - 25 + 21 = 0$$

Simplifying, I have: $x^2 - 10x + 25 - 4 = 0$
$$\text{or}$$
$$(x - 5)^2 - 4 = 0$$

Now we factor: $\{(x - 5) - 2\}\{(x - 5) + 2\} = 0$
Simplifying again: $\{x - 7\}\{x - 3\} = 0$

This equation gives us two options: $0(x - 3) = 0$ or $(x - 7)0 = 0$
So now we see that we must solve the two equations $x - 7 = 0$ and $x - 3 = 0$.
Finally, we see that $x = 7$ or $x = 3$.

72

Mentor: Just so! Caroline, how would you describe the several steps we must go through without using any numbers?

Caroline: I) Given a trinomial expression, we consider the squared term and the term with just a multiple of x as the start of a perfect square in need of a last constant term.

II) We add the needed constant to complete the square and then subtract it out to keep the equation balanced.

III) We combine the two constant terms, and

IV) Treat the constant term as a square.

V) Finally, we can factor the whole expression as the difference of two squares.

VI) From this point, we can set each of the factors equal to zero and solve the simpler linear equations.

Between the Scenes

1) Solve each of the following equations by completing the square.

i) $x^2 + 2x - 8 = 0$ ii) $x^2 - 2x - 8 = 0$ iii) $x^2 + 2x - 15 = 0$ iv) $x^2 - 2x - 15 = 0$

v) $x^2 + 14x + 45 = 0$ vi) $x^2 - 14x + 45 = 0$

2) Solve the equation with a and b being some numbers: $x^2 + 2ax + a^2 - b^2 = 0$.

Act 5   Scene 6   Solving a Quadratic Equation II

Mentor: In the previous scene, we have learned how to complete the square and factor the difference of two squares, so let's complicate the problem again: $x^2 + 3x - 7 = 0$.
Sam: First we write the equation with a multiple of 2 as the coefficient of the x term.

$$x^2 + 2\left(\frac{3}{2}\right)x - 7 = 0$$

Next we complete the square by adding, and subtracting, $\left(\frac{3}{2}\right)^2$:

$$x^2 + 2\left(\frac{3}{2}\right)x + \left(\frac{3}{2}\right)^2 - \left(\frac{3}{2}\right)^2 - 7 = 0$$

Now I want to write the first three terms as a square.

$$(x + \frac{3}{2})^2 - \left(\frac{3}{2}\right)^2 - 7 = 0$$

Here we want to start making the last two terms into a square, factoring out the minus.

$$(x + \frac{3}{2})^2 - (\left(\frac{3}{2}\right)^2 + 7) = 0$$

At this point, we make a single fraction of the last two terms.

$$(x + \frac{3}{2})^2 - (\frac{3^2}{2^2} + 7\times\frac{2^2}{2^2}) = 0$$

We can now combine the last two fractions as one fraction.

$$(x + \frac{3}{2})^2 - (\frac{3^2 + 2^2\times7}{2^2}) = 0$$

If at this juncture we want to see the original -7, we will have to change two signs in the numerator of the last term.

$$(x + \frac{3}{2})^2 - (\frac{3^2 - 2^2\times(-7)}{2^2}) = 0$$

Because we have to make our final expression the difference of two squares, we will square the square root of the last term.

$$(x + \frac{3}{2})^2 - \left(\sqrt{\frac{3^2 - 2^2\times(-7)}{2^2}}\right)^2 = 0$$

Here we can factor the difference of two squares.

$$\left[(x + \frac{3}{2}) - \sqrt{\frac{3^2 - 2^2\times(-7)}{2^2}}\right] \times \left[(x + \frac{3}{2}) + \sqrt{\frac{3^2 - 2^2\times(-7)}{2^2}}\right] = 0$$

Sam: We have a product equal to zero, so we know that one or the other factor equals zero.

$$\left(x + \frac{3}{2}\right) - \sqrt{\frac{3^2 - 2^2 \times (-7)}{2^2}} = 0$$

or

$$\left(x + \frac{3}{2}\right) + \sqrt{\frac{3^2 - 2^2 \times (-7)}{2^2}} = 0$$

Thus giving us the two solutions.

$$x = -\frac{3}{2} + \sqrt{\frac{3^2 - 2^2 \times (-7)}{2^2}}$$

or

$$x = -\frac{3}{2} - \sqrt{\frac{3^2 - 2^2 \times (-7)}{2^2}}$$

We can further simplify this expression by noting that we have a common denominator of 2 if we remove the 2^2 from the denominator of the radical.

$$x = \frac{-3 + \sqrt{3^2 - 2^2 \times (-7)}}{2}$$

or

$$x = \frac{-3 - \sqrt{3^2 - 2^2 \times (-7)}}{2}$$

Mentor: What an effort! Now replace the constants 3 and 7 by the letters 'b' and 'c' respectively and repeat the exercise.

Between the Scenes

Solve: 1) $x^2 + 6x - 5 = 0$ 2) $x^2 - 3x - 11 = 0$ 3) $x^2 + 7x - 6 = 0$
4) $x^2 - 5x + 1 = 0$ 5) $x^2 + 8x + 4 = 0$ 6) $x^2 - 2x - 13 = 0$

Assuming that your equation is $ax^2 + bx + c = 0$, where $a \neq 0$, show the steps required to reach the standard solution to the equation to be: $x = \frac{-b \pm \sqrt{(b^2 - 4ac)}}{2a}$.

Act 5   Scene 7   Pascal's Triangle

Mentor: We have learned how to multiply two binomials, say $(x + 3)(x + 4)$, by multiplying each term in the second by each term in the first, giving us $x^2 + 4x + 3x + 3 \times 4$, or $x^2 + 7x + 12$. In this scene, we will look for a pattern found by multiplying a simple binomial by itself a number of times. We look for the transitions as we multiply the last result by the factor. What is the expanded version of $(x + 1)^2$?

Dmitri: $(x + 1)^2$ means $(x + 1)(x + 1)$, so with 2 factors of $(x + 1)$.

$$(x + 1)(x + 1) = x^2 + 2x + 1$$

Mentor: That was easy, but what about $(x + 1)^3$?

Dmitri: I would multiply $(x + 1)(x^2 + 2x + 1)$ as $x(x^2 + 2x + 1) + 1(x^2 + 2x + 1)$, giving me $x^3 + 2x^2 + x + x^2 + 2x + 1$. Simplifying, we get $x^3 + 3x^2 + 3x + 1$.

Mentor: At this juncture, we have enough information to develop our pattern, but two simple cases could help us by providing the earlier parts of the pattern. What does $(x + 1)^1$ mean?

Matt: $(x + 1)^1$ means we have one factor of $(x + 1)$, so $(x + 1)^1 = x + 1$.

Mentor: Excellent. Now what do we do with $(x + 1)^0$?

Matt: How about 0?

Mentor: Let us look at a fraction such as $\frac{9}{9}$, writing it in two ways.

Matt: 9 goes into 9 one time, so $\frac{9}{9} = 1$.

Mentor: Good. We have done half of the problem. We see that $9 = 3^2$, so we can look at the exponents in this expression. What does $\frac{3^2}{3^2}$ mean? If we add exponents when we multiply, what do we do with the exponents when we divide?

Matt: If we subtracted, we would get 3^0, or 1. This means that $(x + 1)^0 = 1$.

Mentor: Excellent. So let us arrange these results from the smallest exponent, 0, at the top, to the largest exponent expression in the bottom. Can you see at least one pattern?

$$1$$
$$x + 1$$
$$x^2 + 2x + 1$$
$$x^3 + 3x^2 + 3x + 1$$

Dmitri: Clearly there is a line of 1's running down the right side of the triangle.

Mentor: Yes. What else?

Dmitri: All the coefficients of the highest powers of x, running down the left side of the triangle, are also 1.

Mentor: Let's calculate another row to help see yet another pattern.
$$(x + 1)(x^3 + 3x^2 + 3x + 1) = ?$$

Dmitri: We multiply each term of the second polynomial by x and then by 1, and add.

$(x + 1)(x^3 + 3x^2 + 3x + 1) = x(x^3 + 3x^2 + 3x + 1) + 1(x^3 + 3x^2 + 3x + 1)$, or
$x^4 + 3x^3 + 3x^2 + x + x^3 + 3x^2 + 3x + 1 = x^4 + 4x^3 + 6x^2 + 4x + 1$.

76

$$1$$
$$x + 1$$
$$x^2 + 2x + 1$$
$$x^3 + 3x^2 + 3x + 1$$
$$x^4 + 4x^3 + 6x^2 + 4x + 1$$

Mentor: We continue to see the 1's running down the right side and the increasing powers of x running down the left side, but what about the middle?

Sam: I see the 2 of 2x on the second line being the sum of the coefficient of the above x and the 1 on the right. This sum idea works on the next pair of lines as well. The 3 of $3x^2$ can be seen as the sum of the 1 in x^2 and the 2 in 2x on the line above.

Mentor: That pattern works on the left. What is happening on the right side of the triangle?

Sam: The pattern is symmetrical about a vertical line starting at the 1 on the top and going all the way down.

Mentor: Do we have to multiply the last line by (x + 1) to get the next line, or can we follow our pattern?

Sam: I would add the 1 from x^4 and the 4 from the $4x^3$ to get $5x^4$. Of course the first term is x^5, with the coefficient of 1 and the exponent of 5. Following this pattern, we have the next row.

$$1$$
$$x + 1$$
$$x^2 + 2x + 1$$
$$x^3 + 3x^2 + 3x + 1$$
$$x^4 + 4x^3 + 6x^2 + 4x + 1$$
$$x^5 + 5x^4 + 10x^3 + 10x^2 + 5x + 1$$

Mentor: How do we <u>find the coefficients on the next line</u>?

Sam: We <u>add the coefficients of the two terms on the line above that are to the left and to the right of our current term.</u>

Mentor: Fill in the next line.

Dmitri: The next line is $x^6 + 6x^5 + 15x^4 + 20x^3 + 15x^2 + 6x + 1$.

$$1$$
$$x + 1$$
$$x^2 + 2x + 1$$
$$x^3 + 3x^2 + 3x + 1$$
$$x^4 + 4x^3 + 6x^2 + 4x + 1$$
$$x^5 + 5x^4 + 10x^3 + 10x^2 + 5x + 1$$
$$x^6 + 6x^5 + 15x^4 + 20x^3 + 15x^2 + 6x + 1$$

Mentor: Any other patterns?

Dmitri: The coefficients running down left and right sides in the second row count up by ones: 1, 2, 3, … , 6.

Mentor: Later on, we will find this triangle useful for finding formulas for sums.

Between the Scenes

Explain why $(3a + 5b)^3 = 27a^3 + 135a^2b + 225ab^2 + 125b^3$.
Multiply out and explain the sign pattern in each of the following.

      1) $(3a - 5b)^3$
      2) $(4x + 7y)^3$
      3) $(4x - 7y)^3$
      4) $(2x + 9y)^4$

Act 5   Scene 8   Solving a System: Quadratic & Linear

Mentor: We have drawn lines and found where they intersect and we have drawn parabolas. Now we will find out where a line and a parabola intersect. Here is a system of two equations as shown in Figure 61. Solve for their points of intersection, if any exist.

$$y = 2x - 5 \qquad (2xm5)$$
$$y = x^2 - 7x + 9 \qquad (x2m7xp9)$$

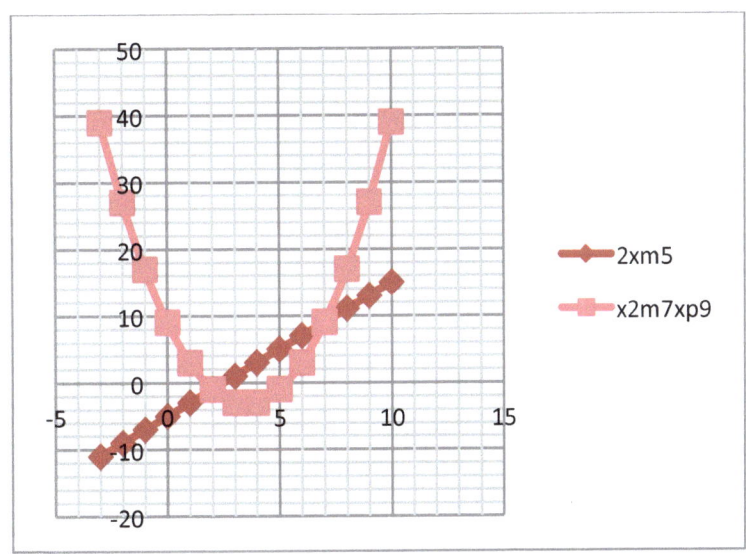

Figure 61

Madeleine: Where the two lines intersect the values for x and y must be equal, so I will substitute the 2x – 5 in the first equation for the y in the second equation. Then I will have only one variable, x.

$$2x - 5 = x^2 - 7x + 9$$

If I subtract 2x from both sides of the equation, all the x terms will be on the right.

$$-5 = x^2 - 9x + 9$$

Then I will add 5 to both sides, so I will have a quadratic expression in x equal to zero.

$$0 = x^2 - 9x + 14$$

Mentor: How does all this help you?
Madeleine: I can solve this equation two different ways. The simplest only requires that I factor the quadratic expression on the right. I see that $x^2 - 9x + 14 = (x - 2)(x - 7)$. Now if $(x - 2)(x - 7) = 0$, then either x – 2 = 0 or x – 7 = 0.
Mentor: What reason do you have for this conclusion?
Madeleine: <u>If the product of two numbers equals zero, then one of them equals zero.</u>

79

Mentor: Excellent. What then?

Madeleine: $x - 2 = 0$ means that $x = 2$, and $x - 7 = 0$ means that $x = 7$ by adding a 2 or a 7 to both sides of the appropriate equation.

Mentor: Is that the complete answer?

Madeleine: No. I have to find the y that goes with each x: if $x = 2$ then $y = -1$, and if $x = 7$, then $y = 9$.

Mentor: Where do the two curves intersect?

Madeleine: At A(2, -1) and B(7, 9).

Between the Scenes

Draw the graphs of each pair of curves and find the intersections of the following systems.

1) $y = -x + 4$; $y = x^2 - 6x - 2$
2) $y = 2x + 8$; $y = x^2 + x - 12$
3) $y = 5x - 9$; $y = x^2 + 4x - 21$
4) $y = -3x + 10$; $y = x^2 - 8x - 14$

Act 6   **Graphing Cubic and Higher Polynomials**

Scene 1   **Graphing Factored Cubic Polynomials**

Mentor: We say that x^2, $(x-1)^2$, and $(x-2)(x+5)$ are all quadratic polynomials, because multiplying each expression to remove the parentheses creates one term with x^2 in it. The highest exponent tells us the <u>degree of the polynomial</u>. So we say that the degree of each of the polynomial expressions x^3, $(x-1)^2(x+5)$, and $(x-6)(x+2)(x-10)$ is of degree 3. If we multiply a polynomial of degree 1 times a polynomial of degree 2, what will the degree of the product polynomial be?

Paul: The degree of the new polynomial will be 3, because we <u>add the exponents when we multiply the factors</u>. This is just like multiplying $3^2 \times 3^5 = 3^7$.

Mentor: Now tell us what the shape of the curve $y = (x-1)^2(x+5)$ will be.

Paul: It will be shaped like the cubic $y = x^3$, except that it will have some bumps that $y = x^3$ does not. Here are the graphs of these curves.

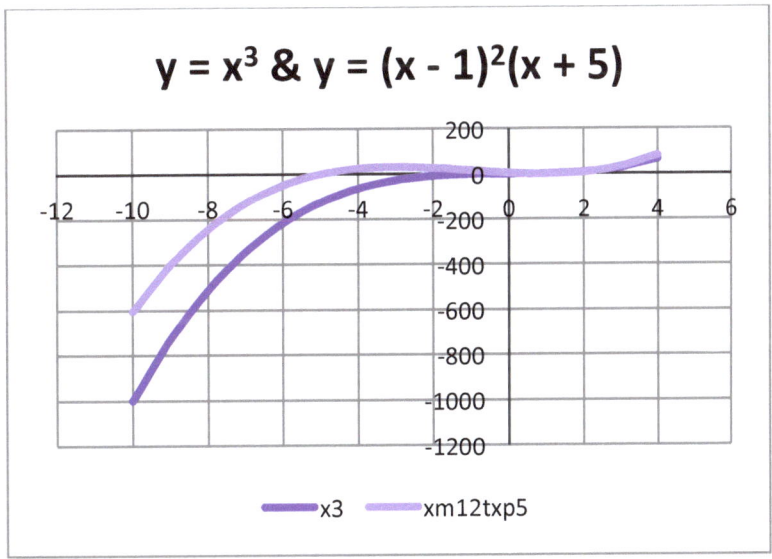

Figure 62

Mentor: What will be the degree of the polynomial in $y = (x-6)(x+2)(x-10)$?

Paul: Because each factor is of first degree, we add the understood-but-not-written exponents to get $1 + 1 + 1 = 3$.

Mentor: What will the shape of this polynomial curve be?

Paul: Its shape will be the same as those curves in Figure 62, except that the curve has zeros at -2, 6, and 10, where it passes through the x-axis. Each cubic curve will go up on the right side and down on the left side with some mountains and valleys in between. Here is its graph in Figure 63.

81

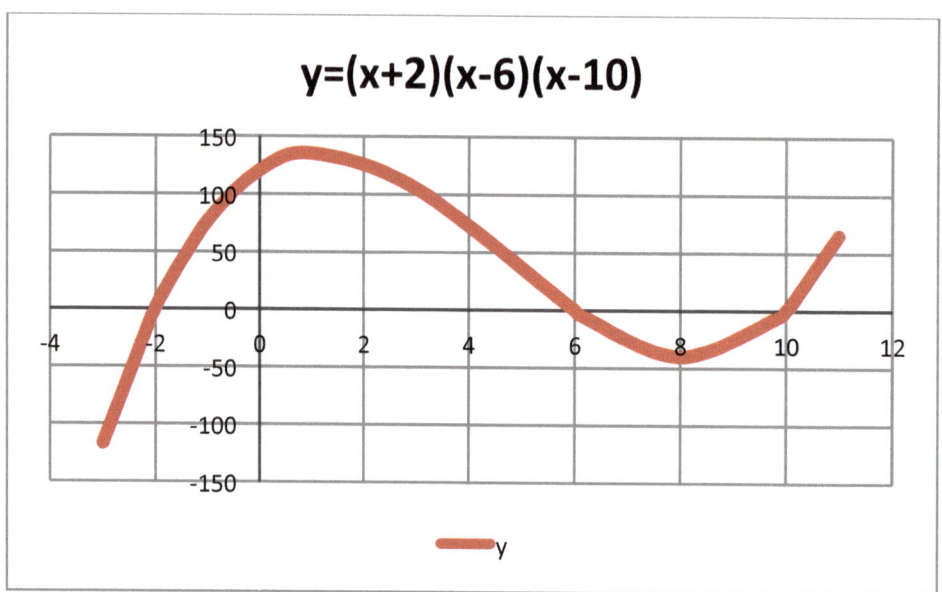

Figure 63

Mentor: How would you compare the graphs of $y = x^3$, "x3", and $y = (x - 5)^3$, "xm53"?
Paul: The graph of the second curve is the same as that of the first but moved to the right 5 units. Figure 64 below shows the comparison.

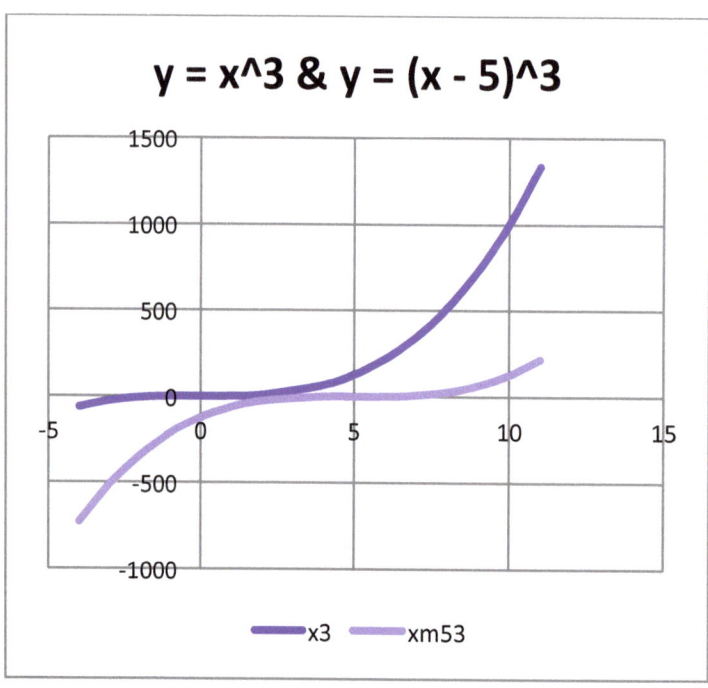

Figure 64

Between the Scenes
Graph the following cubic curves.

    1) $y = x(x - 1)(x - 2)$
    2) $y = x^2(x - 5)$
    3) $y = x(x - 5)^2$
    4) $y = (x - 5)^3$

Act 6   Scene 2   Drawing Higher Degree Factored Polynomials

Mentor: We have seen that quadratic curves look like a big U when the coefficient of the x^2 term is positive, as in $y = 3x^2 - 8$, and an upside down U when the coefficient of x^2 is negative, as in $y = -3x^2 + 7$.   By contrast we have seen the cubic curves, such as $y = 4x^3 + 5x$, look like a bent straight line, such as $y = 3x + 1$, reaching from the third quadrant to the first, whereas the cubic curve with a negative coefficient of the x^3 term, as in $y = -2x^3 + 8x^2$, comes down from the second quadrant into the fourth quadrant, much like the line $y = -2x + 9$. In this scene we will explore the behavior of higher degree polynomials. What does the curve

$$y = (x - 3)^2 (x - 5)^2$$

look like? What is its degree?

Madeleine: I will start with the degree. The sum of the degrees of the two factors is 4, so that is its degree. Its shape will be a U like $y = 3x^2 - 8$. We also know that $y = 0$ at two places on the curve, at $x = 3$ and at $x = 5$.

Mentor: Excellent. Can you draw the graphs of each of the curves $y = (x - 3)^2$ (i.e. on graph: xm32) and $y = (x - 5)^2$ (i.e. on graph: xm52)?

Madeleine: Yes. That is easy!

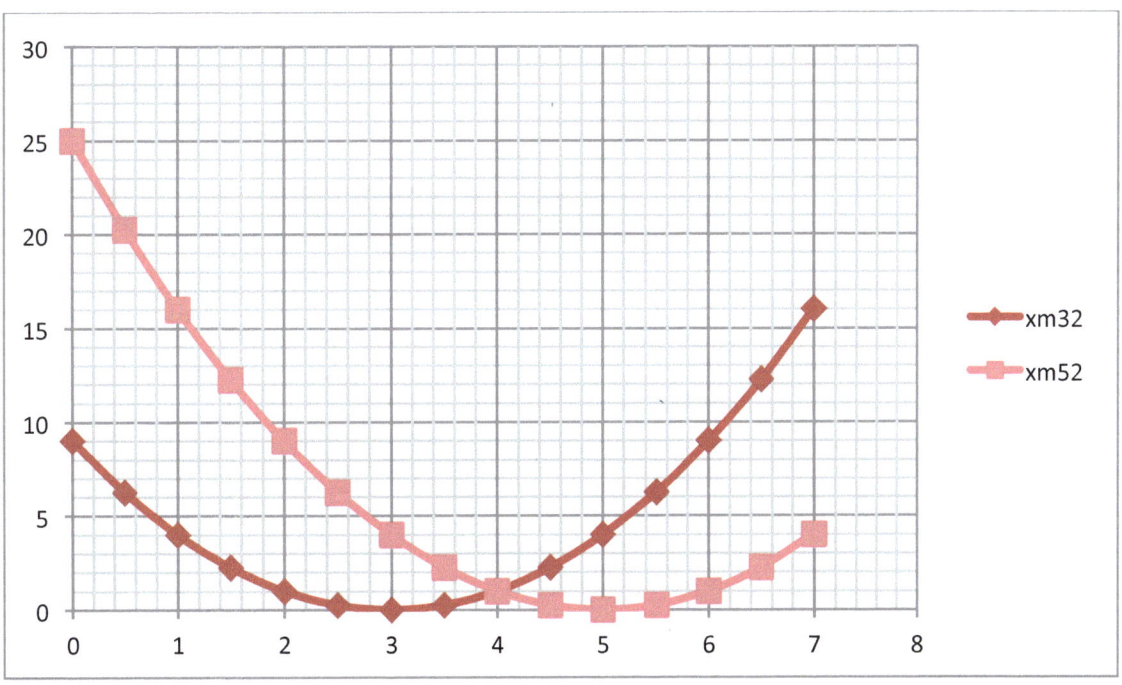

Figure 65

Mentor: What can we say about these two curves and the x-axis?

Madeleine: Each of these curves is above the x-axis except at $x = 3$ and $x = 5$ where $y = 0$.

Mentor: What can you say about the product of these two curves?

Madeleine: The entire curve, except for these two points, will be above the x-axis.

Mentor: What would this product curve look like?

83

Madeleine: The two points (3, 0) and (5, 0) will belong to the new curve, because the product of zero and any other number is zero. For x < 3 and x > 5, the product curve grows larger the farther we go away from the two zeros.

Mentor: What happens in between x = 3 and x = 5?

Madeleine: The product is positive, so there is a little hill between the two zeros. Here is my picture in Figure 66. (NB: 'xm32' is $y = (x - 3)^2$ and 'xm52' is $y = (x - 5)^2$)

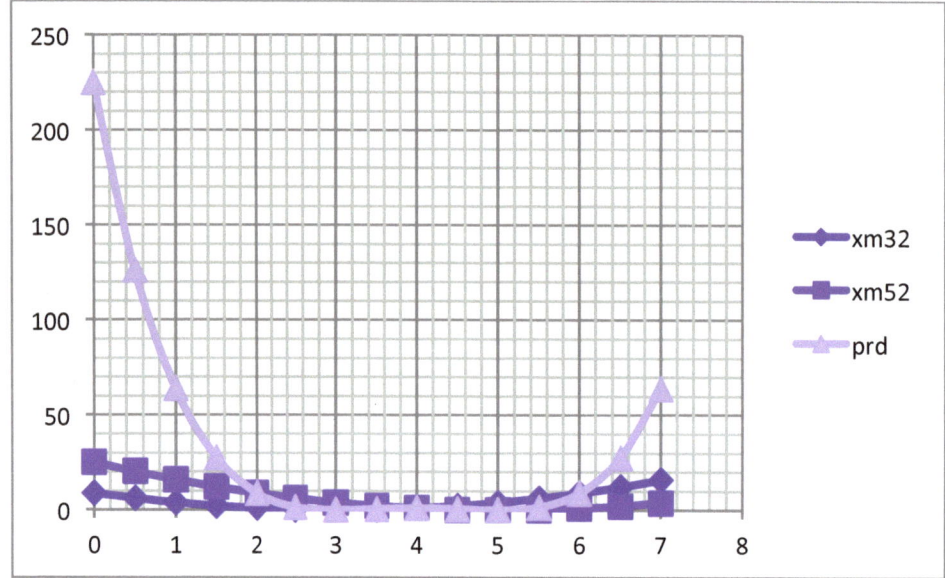

Figure 66

Mentor: A very "little hill!" What could we do to see the middle between the zeros?

Madeleine: In Figure 67, you can see the factors in Series1 and Series2, and the product in Series3.

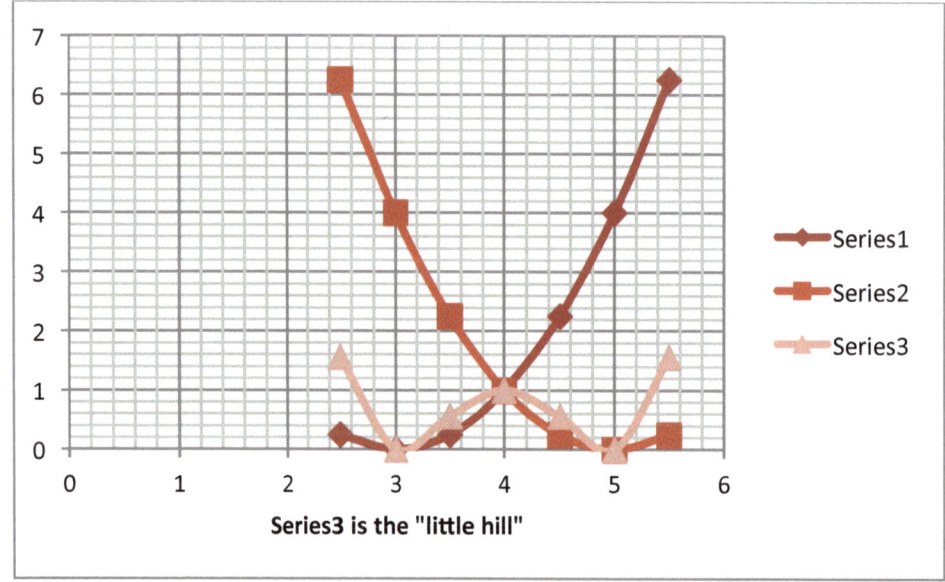

Figure 67

Mentor: Figure 67 gives us a good idea of product between the two zeros. Now what will the negative of the product look like?

84

Madeleine: The negative will be the mirror image through the x-axis.

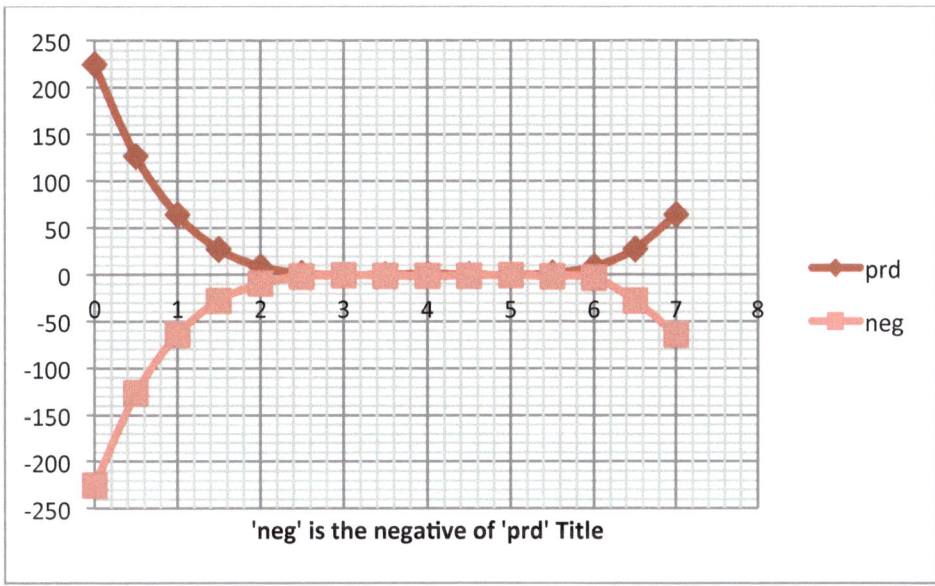

Figure 68

The graph of "prd" is that of our product, and "neg" is its negative.
Mentor: Exactly. Now what about the reciprocal?
Madeleine: Where the product is far away from the x-axis, its reciprocal is near the x-axis. Also, where the product is close to the x-axis, its reciprocal is far away from the x-axis.

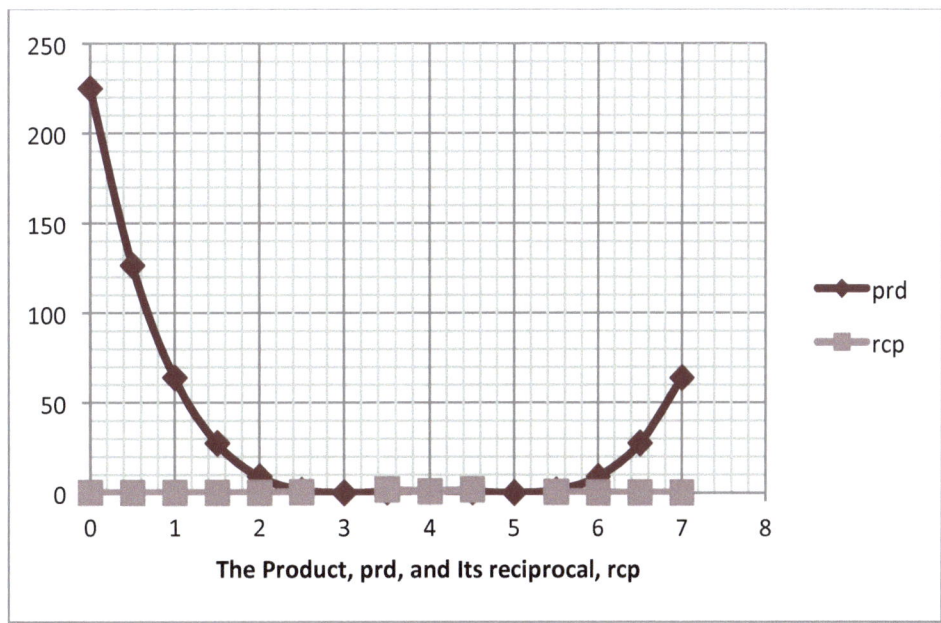

Figure 69

Mentor: What is wrong with this chart, Figure 69?

85

Madeleine: It is incomplete! I figured it out. Close to the x-coordinates 3 and 5 are the small numbers in the product. So close to the x-coordinates 3 and 5 we will find large reciprocals. However, we did not put those into the graph! Try again!

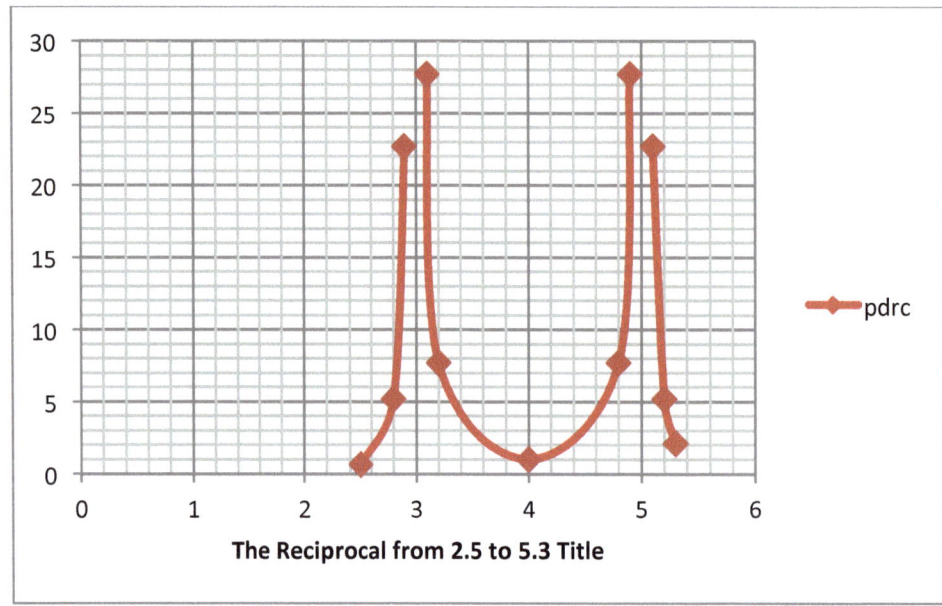

Figure 70

Mentor: Now I can see where the problem was!

Between the Scenes
For each of the polynomial curves below, after stating the degree of the product,

     a) Graph each factor with the other factors,
     b) Graph the product of the factors,
     c) Graph the negative of the product, and
     d) Graph the reciprocal of the product.

1. $y = (x + 8)^2(x + 4)^3(x - 1)^4$
2. $y = (x - 2)^3(x - 7)^3(x - 10)^3$
3. $y = (x + 10)^4(x + 6)^3(x + 2)^2(x - 3)(x - 8)$

Act 7   Newton's Addition: Polynomial Differentiation

Scene 1   A Secant and Tangent To a Parabola

Mentor: In this act we seek to trace the path of a (Stomp) rocket, indicating its height when gravity stops its upward flight and finding its speed at any given time during the flight. Alternatively, picture a hammer thrower, spinning around, letting go at a certain moment, and tracing the straight path of the "hammer" after the release. What is the slope of the line the hammer travels on? We will use the slope of the secant in Figure 71 to approximate our tangent line.

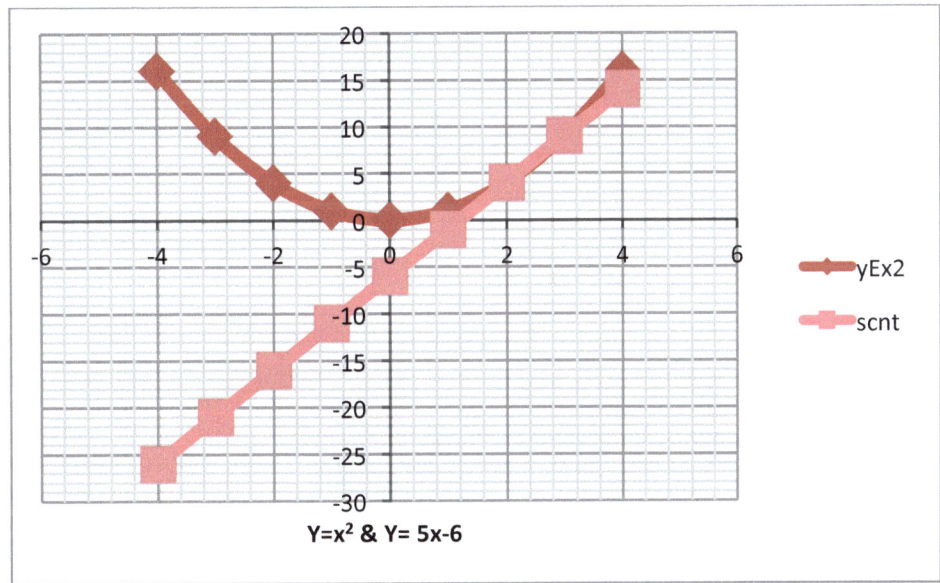

Figure 71

I want you to consider the hammer starting on the parabola on the origin before being released at the point (2,4). Why can't we just write the equation of the tangent line?
Emma: We only have one point, (2,4), so we cannot calculate the slope.
Mentor: If the line went through the point (3,9), could you write the equation of the line?
Emma: Sure! Calculating the slope of any line through the points (x, y) and (2, 4) would give the same slope as that of points (3, 9) and (2, 4).

$$\frac{y-4}{x-2} = \frac{9-4}{3-2}$$

Simplifying, we have

$$y - 4 = 5(x - 2)$$
$$\text{or}$$
$$y = 5x - 6$$

Mentor: Excellent. The problem we have now is dealing with the knowledge that the hammer will not cut in inside the parabola before departing on its course. How can we write the equation of a line that is almost the tangent line, cutting across less of the inside of the parabola?

87

Trix: Instead of (3, 9), we should pick (2.1, 4.41), the hammer cuts across just one tenth of the distance inside the parabola.

Mentor: What would the equation be?

Trix: We would have the same first side of the equation.

$$\frac{y-4}{x-2} = \frac{4.41-4}{2.1-2}$$

Simplifying,

$$\frac{y-4}{x-2} = \frac{0.41}{0.1}$$

or

$$\frac{y-4}{x-2} = \frac{4.1}{1}$$

We now see that the slope is 4.1 as in Figure 72.

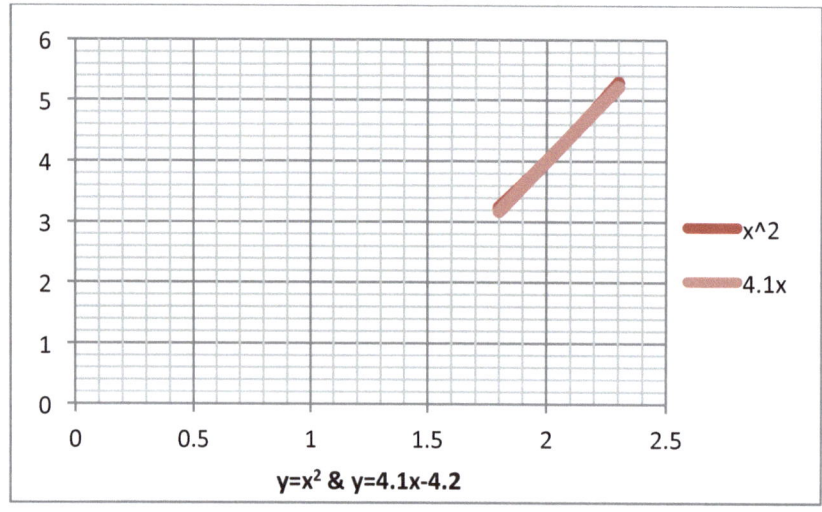

Figure 72

Mentor: So what is the problem with our new graph?

Maddy: The parabola and the secant line seem to be the same set of points, even though we know they are not!

Mentor: The graphs may not be different in Excel, so we can look at the numbers alone. What calculation would give us a better approximate slope of the tangent line?

Maddy: Let's try using (2.01, 4.0401) as the second point.

$$\frac{y-4}{x-2} = \frac{4.0401-4}{2.01-2}$$

or

$$\frac{y-4}{x-2} = \frac{0.0401}{0.01}$$

This gives us the slope 4.01, much closer to the number 4.

Mentor: What would happen if we used (2.001, 4.004001)?
Ursula: I think we would find the new slope even closer to 4.

$$\frac{y-4}{x-2} = \frac{4.00040001 - 4}{2.0001 - 2}$$

$$\frac{y-4}{x-2} = \frac{0.00040001}{0.0001}$$

Now the slope will be 4.0001. So each time we insert another zero before the 1, we find the slope that much closer to 4. So I think the tangent line will have 4 as its slope and the hammer, or ball, will travel along the line y = 4x – 4.
Mentor: Can anybody find a rule or pattern for the tangent line to the parabola at any point, say (2, 4)?
Georgia: It looks like the y-coordinate to me!
Mentor: Fine. So what do you expect the slope of the tangent line at (3, 9) to be?
Georgia: 9.
Mentor: OK. Let's calculate a secant line and its slope. What do you choose to do?
Georgia: My second point is (3.001, 9.006001), and the equation would be:

$$\frac{y-9}{x-3} = \frac{9.006001 - 9}{3.001 - 3}$$
giving us

$$\frac{y-9}{x-3} = \frac{0.006001}{0.001}$$

So now we see that the secant slope will be 6.001, making the tangent slope 6.
Mentor: Do you want to adjust your initial guess about the slope of the tangent line?
Georgia: Oh, yes! In both cases, the slope is twice the x-coordinate.
Mentor: So what should the slope of the tangent line at (5, 25) be?
Madeleine: It will be 10.
Mentor: How are you going to convince us?
Madeleine: I will pick a point very close to 5, like 5.001, and calculate the slope of the secant line.

$$\frac{y-25}{x-5} = \frac{25.010001 - 25}{5.001 - 5}$$

$$\frac{y-25}{x-5} = \frac{0.010001}{0.001}$$

$$\frac{y-25}{x-5} = \frac{10.001}{1}$$

So, I was right! For a secant line through a point close to (5, 25) the slope will be twice the x-coordinate, namely 10. More generally, the slope of the tangent line at any point on y = x² is 2x.

Mentor: I would like to introduce a different approach. Let's go back to our first equation of one slope equal to another.

$$\frac{y-4}{x-2} = \frac{9-4}{3-2}$$

On the right side, we see that the numerator is the difference of two squares: $9-4$ or $3^2 - 2^2$. What can we do with the difference of two squares?

Piper: I can factor it to $(3-2)(3+2)$, so the fraction on the right reads:

$$\frac{9-4}{3-2} = \frac{(3-2)(3+2)}{(3-2)}$$

Now in this factored form, we see that the fraction reduces to $(3+2) = 5$. This gives us the same slope as Emma found when we started.

Mentor: What can you do with Trix's example, the slope though $(2, 4)$ and $(2.1, 4.41)$?

$$\frac{y-4}{x-2} = \frac{4.41-4}{2.1-2}$$

Piper: We can see that 4.41 is the square of 2.1, so we have the difference of two squares in the second fraction.

$$\frac{4.41-4}{2.1-2} = \frac{(2.1-2)(2.1+2)}{2.1-2}$$

Now the $2.1 - 2$ reduces in the second fraction, leaving us $(2.1 + 2) = 4.1$ as the slope.

Mentor: How could we have approached Madeleine's tangent at $(5, 25)$?

Piper: We now know that $(25.010001 - 25)$ is a difference of two squares, so we can factor the numerator.

$$\frac{25.010001 - 25}{5.001 - 5} = \frac{(5.001 - 5)(5.001 + 5)}{5.001 - 5}$$

Of course, this new fraction can be reduced to just $5.001 + 5 = 10.001$. This was exactly Madeleine's solution!

Mentor: So what have we just observed about calculating slopes of secant lines?

Piper: We have two methods of calculating the slope of a secant line: (1) subtract in numerator and denominator and then divide or (2) factor into the difference of two squares, reduce, and add.

Maddy: So I could have simplified my fraction by factoring and reducing.

$$\frac{4.0401 - 4}{2.01 - 2} = \frac{(2.01 - 2)(2.01 + 2)}{2.01 - 2}$$

or

$$2.01 + 2 = 4.01.$$

Mentor: Exactly. You now have two different methods of finding the slope.

Act 7   Scene 2   A Secant and Tangent to a Cubic Curve

Mentor: We have seen that we can calculate the slope of a secant line to a parabola in two ways. Do our two methods work with a cubic and do we find that the rule for the slope of the tangent line, that is doubling the argument, is the same as with the quadratic? Let's try $y = x^3$ at $(2, 8)$

Emma: I am going to take a point close to the given point $(2, 8)$ and calculate the slope of the secant line.

$$\frac{y-8}{x-2} = \frac{8.012006001 - 8}{2.001 - 2}$$

Subtracting first, we have the new expression:

$$\frac{y-8}{x-2} = \frac{0.012006001}{0.001}$$

If we multiply numerator and denominator by 1000, we acquire the slope 12.006001. If we look at this numerator as the difference of two cubes, then we have the following fraction.

$$\frac{2.001^3 - 2^3}{2.001 - 2} = \frac{(2.001 - 2)(2.001 \times 2.001 + 2.001 \times 2 + 2 \times 2)}{2.001 - 2}$$

At this juncture, we see that the common factor of $(2.001 - 2)$ can be reduced, leaving us with the sum of the three remaining terms in the numerator:

$$2.001 \times 2.001 + 2.001 \times 2 + 2 \times 2 = 4.004001 + 4.002 + 4, \text{ or}$$
$$12.006001$$

Mentor: What do you think is the slope of the tangent line at $(2, 8)$? Why?

Emma: The slope is 12, because the more zeros between the 2 and the decimal 1, the closer to 12 the number will become.

Mentor: Does Madeleine's rule still hold here, namely double the x-coordinate to get the slope?

Emma: No. We have to multiply the x-coordinate by 6 to get to 12. The rule seems to be different for cubic polynomials.

Mentor: When you use the factoring method, how many terms are left for the final addition?

Emma: Three and each term is about 3×4, where 4 is either the sum of two 2's or the product of two 2's.

Mentor: Let's look at $y = x^3$ at the point $(5, 125)$. What is your guess about the slope of the tangent line at that point?

Ursula:

$$\frac{5.001^3 - 5^3}{5.001 - 5} = \frac{(5.001 - 5)(5.001^2 + 5.001 \times 5 + 5 \times 5)}{5.001 - 5}$$

After we reduce the fraction by the common $(5.001 - 5)$ we are left with the trinomial

$$5.001^2 + 5.001 \times 5 + 5 \times 5 = 25.005001 + 25.005 + 25$$

This is all equal to 75.010001, which I suspect is 3×5^2.

Mentor: This sounds good to me, but how did you get from $y = x^3$ at (5, 125) to 3×5^2?

Ursula: I'm not sure, but the coefficient 3 is the exponent of the monomial x^3, and the 5^2 is the x-coordinate squared.

Mentor: Nice. Would your rule work at the point (2, 8)?

Ursula: Yes! $12 = 3 \times 2^2$.

Mentor: So, when we calculate the slope of the tangent line at a point on the cubic curve, what do we do to get the slope?

Ursula: We make the original exponent the coefficient of x and subtract 1 from the original exponent to get the new exponent, giving us a formula for the slope at that point. So we can say, the slope of the tangent line at any point (x, x^3) will be $3x^2$.

Mentor: We are making progress. There are two forms of notation you should be aware of before going on to the next Scene.

The first item on the list is one way we express the slope of the tangent line to a curve: $y' = 3x^2$ in the example Ursula explained. " y' " we call the derivative of y at any point (x, y). You will see this expression used often in the rest of this text.

The second item of importance is expressed by the word 'function.' The idea of a function is straight forward, but may take some time to get used to. We have talked about "curves" in this text to avoid more new notation, but you should have even this brief introduction. We name a function, 'f' or some other letter in the middle of the alphabet, usually g or h. If x is an argument of a curve and y is the value found with x in the ordered pair (x, y), we will say f(x) = y (Read "f of x equals y") or in an example, f(x) = $3x^2$. Our definition of a function states that every argument has exactly one value. A circle $x^2 + y^2 = 4$ can have two values for the argument x = 0, namely y = 2 and y = -2, so this is not a function.

Between the Scenes

Apply Ursula's rule to each of the points on $y = x^3$, and show that it works, or does not, by approximating the tangent's slope.

    1)   (3, 27)
    2)   (4, 64)
    3)   (7, 343)
    4)   (-2, -8)

Act 7   **Scene 3**   **Tangent to a Quartic Monomial**

Mentor: Now we want to extend our pattern to cover the fourth degree monomials, $y = x^4$. How should we start?

Georgia: Take a point on the curve and calculate the slopes of secants, each with a second point closer to the original point.

Mentor: What two points do you think we should use?

Georgia: To keep the numbers as small as reasonable, (2, 16) and (3, 81).

$$\frac{y - 16}{x - 2} = \frac{81 - 16}{3 - 2}$$

So the slope of this secant equals 65.

Mentor: What would your next second point be?

Georgia: (2.01, 16.3224). This gives me the slope of $\frac{16.3224 - 16}{2.01 - 2} = \frac{0.3224}{0.01}$, or 32.24.

Mentor: Are you prepared to say what the slope is now?

Georgia: I think it will be 32. If $y = x^4$, then the slope $y' = 4x^3$, or in this case $4(2^3) = 4 \times 8 = 32$.

Mentor: You seem to have this under control. I would like to have you set the slope up in the factored form much as we did in the last scene.

Georgia: This way I would have the slope, m, equal to this fraction.

$$m = \frac{2.01^4 - 2^4}{2.01 - 2}$$

Or,

$$\frac{(2.01 - 2)(2.01^3 + 2.01^2 \times 2 + 2.01 \times 2^2 + 2^3)}{2.01 - 2}$$

Reduced: $2.01^3 + 2.01^2 \times 2 + 2.01 \times 2^2 + 2^3$

And simplified:

$$32.2408$$

This looks like about the same number we got before!

Mentor: How many terms do we have in the next-to-last expression?

Georgia: Four. I see that the derivative has as its coefficient the old exponent, that is the number of terms in the reduced expression above multiplied by the x-coordinate of the point of tangency raised to the third or one less than the original exponent. We have:

$$y' = 4x^3$$

where x = 2.

Mentor: What do you think the derivative of $y = x^5$ will be?

Georgia: $y' = 5x^4$.

Mentor: Once again, you are on target! Looking for the pattern brings us to using algebra. In this manner, we are going to see a single letter instead of a number with many digits. So, how will we factor $x^4 - a^4$?

Georgia: I know that I want one factor to be $(x - a)$ and the other has to have an x^3 and an a^3 to have the product be $x^4 - a^4$.

Mentor: If I wrote $(x - a)(x^3 + q + a^3)$, with q being the questionable material, would that satisfy your idea?

Georgia: Yes. And now I will multiply $(x - a)$ times the new factor $(x^3 + q + a^3)$.

$$x^4 - a^4 = (x - a)(x^3 + q + a^3),$$
$$\text{or}$$
$$= x^3(x - a) + q(x - a) + a^3(x - a),$$

and removing several pairs of parentheses,

$$= x^4 - ax^3 + q(x - a) + a^3x - a^4.$$

Mentor: What is the value of everything between the x^4 and $- a^4$?
Georgia: It has to be zero. Solving for q then, gives us:

$$-ax^3 + q(x - a) + a^3x = 0,$$
$$\text{or}$$
$$q(x - a) = ax^3 - a^3x$$

Mentor: Fine. What do you see in the expression on the right side of the equation?
Georgia: I can factor out the ax!

$$ax^3 - a^3x = ax(x^2 - a^2),$$
$$\text{and again}$$
$$= ax(x - a)(x + a)$$

So now I can solve the equation:

$$q(x - a) = ax(x - a)(x + a),$$

by dividing both sides by $(x - a)$, giving us

$$q = ax(x + a),$$
$$\text{or}$$
$$ax^2 + a^2x$$

Finally, substituting for q, we have our expression:

$$x^4 - a^4 = (x - a)(x^3 + ax^2 + a^2x + a^3)$$

Mentor: What does this give us for the slope of the secant line?
Georgia: This is the slope:

$$\frac{x^4 - a^4}{x - a} = \frac{(x - a)(x^3 + ax^2 + a^2x + a^3)}{(x - a)},$$
$$\text{reducing}$$
$$= x^3 + ax^2 + a^2x + a^3$$

Mentor: What benefit does this expression have for us?
Georgia: We can see that there are 4 terms, and if we add the degrees of the a's and the x's in each term, we always get 3. Replacing 'a' with an 'x', we have

$$y' = 4x^3$$

Between the Scenes
Calculate the derivatives of the following curves and explain your reasoning:
a) $y = x^5$ b) $y = x^6$ c) $y = x^7$

Act 7   Scene 4   Three Basic Derivatives

Mentor: There are three curves we will examine in this Scene:
1) the constant function y = a,
2) the non-vertical straight line, and
3) the constant multiple of a monomial.

1) What do you think is the derivative of y = 7?
Trix: The derivative is the slope of the tangent line at any particular point on the curve. In this case we can calculate the slope easily:

$$m = \frac{7-7}{3-2}$$

$$= \frac{0}{3-2} \text{ or } 0$$

Mentor: And if we refer to the general point (a, b) on the line y = b?
Trix: Then the slope would be $(b - b)/(x_1 - x_2) = 0$, where the subscripts indicate the two points are different.

2) Mentor: Suppose we consider the line y = 5x – 3. What is its slope?
Trix: We know that the coefficient of the x is the slope of that line, but we can calculate its slope using two points on the line: (a, 5a – 3) and (b, 5b – 3).

$$m = \frac{(5a - 3) - (5b - 3)}{a - b}$$

or

$$= \frac{5a - 3 - 5b + 3}{a - b}$$

Simplifying, we have

$$= \frac{5a - 5b}{a - b}$$

$$= \frac{5(a - b)}{a - b}$$

$$= 5$$

If we used a letter, k, instead of 5, we would have come out the same way with whatever letter we used to indicate the slope.

3) Mentor: What is the derivative of $y = 9x^2$?
Trix: I will set up the equation for the slope of a secant line through $(a, 9a^2)$ and $(b, 9b^2)$.

95

Now the slope of the secant line is m = $(9a^2 - 9b^2)/(a - b)$. Factoring the numerator we have $m = \frac{9(a-b)(a+b)}{a-b}$. On reducing the fraction by the factor $(a - b)$, we are left with $9(a + b)$. Where a approaches b, so they are equal, the slope equals $9(2b)$. Now we recognize that 2b is the derivative of $y = x^2$, where $x = b$.

Mentor: How would you state the rule?

Trix: The <u>derivative of a constant times a function equals the constant times the derivative of the function, or symbolically if f(x) = k g(x), then $f'(x) = k\, g'(x)$.</u>

Between the Scenes

Calculate and explain the derivative of each of the following curves.

a) $y = 3x^4$ b) $f(x) = -8x^5$ c) $y = 17$ d) $f(x) = -2x + 11$

e) $y = 4x^3$ f) $f(x) = -5x^8$ g) $y = 9x - 7$ h) $y = -16$

Act 7   **Scene 5**   **Differentiating Sums and Differences**

Mentor: We have already learned how to differentiate monomials such as $y = 5x^2$ and $y = x^3$. What do you think we will get if we differentiate the polynomial that is the sum of the two?

Piper: I think the <u>derivative of the sum</u>, $y = 5x^2 + x^3$, <u>will be the sum of the derivatives</u>: $y' = 10x + 3x^2$.

Mentor: How would you substantiate your claim?

Piper: I will set the slope of the secant line through $(x, 5x^2 + x^3)$ and $(a, 5a^2 + a^3)$, simplify the fractions and see what happens as x gets close to a.

$$\frac{(5x^2 + x^3) - (5a^2 + a^3)}{x - a} = \frac{5x^2 - 5a^2 + x^3 - a^3}{x - a},$$

factoring the numerator

$$= \frac{5(x^2 - a^2) + (x - a)(x^2 + ax + a^2)}{x - a},$$

and again

$$= \frac{5(x - a)(x + a) + (x - a)(x^2 + ax + a^2)}{x - a},$$

and, yet again

$$= \frac{(x - a) \times [5(x + a) + (x^2 + ax + a^2)]}{x - a}$$

Now simplifying

$$= 5(x + a) + (x^2 + ax + a^2)$$

Now we can let x be equal to a without worrying about dividing by zero.

$$= 5(a + a) + (a^2 + a^2 + a^2)$$

$$= 5 \times 2a + 3a^2$$

So

$$y' = 10\,a + 3a^2$$

Now we have the sum of the two derivatives!

Mentor: What is the slope of the tangent line at the point $(7, 5 \times 7^2 + 7^3)$?

97

Piper: The slope of the tangent line at the point $(7, 5 \times 7^2 + 7^3)$, or $(7, 588)$, will be

$$y'_{(7, 588)} = 5 \times 2 \times 7 + 3 \times 7^2 \text{ or}$$

$$= 70 + 3 \times 49$$

$$= 217$$

That's really steep!

Mentor: It is both steep and correct! Suppose we wanted the derivative of the difference of the two monomials. What is that?

Piper: If $y = 5x^2 - x^3$, then $y' = 10x - 3x^2$. I can show the development by changing the signs on the derivative of the sum.

Between the Scenes

1. Differentiate each of the following polynomials two ways, where $x = 5$ and where $x = a$.
   a) $y = 3x^4 + 4x^5$
   b) $y = 2x^3 - 7x + 4$
   c) $y = 11x^4 - 3x^2 + x$
   d) $y = -16x^2 + 64x + 1$

2. Suppose we factored the x^2 out of $y = 5x^2 + x^3$, differentiating the first factor, x^2, multiplying it by the second factor, $(5 + x)$, and then differentiated the second factor and multiplied it by the original first factor. What do you get? If you just treated $5x^2 + x^3$ as a sum, what would you get. Can you determine what the rule is for differentiating the product of two polynomials? Try differentiating the polynomial in part a) above as a product.

Act 7   Scene 6   Gravity's Effects 0n Motion

Mentor: We have seen a toy rocket shoot straight up in the air, slow down, stop, and return to the ground. The English mathematician and physicist Isaac Newton (1642-1727, British scientist, mathematician, and philosopher who formulated and proved the law of gravity) described the motion as the effect of gravity on an object in flight. If you don't have a toy rocket with an airbag to stomp on, you might try holding a small rock at shoulder height and letting go of it. What happens?

Maddy: It falls to the ground.

Mentor: Why doesn't it go up or off to the side?

Maddy: Gravity pulls it toward the earth.

Mentor: Does the rock travel at the same speed at the bottom of its fall as it did when you let go?

Maddy: Well, I'm a little unsure, but I believe it gains speed as it falls.

Mentor: How could we test your claim: slower at first; faster at the end?

Maddy: I could put my free hand out under the rock very close to my rock-holding hand and feel how hard the rock hits my lower hand. Then I could put my free hand much lower when I drop the rock and feel the force of the falling rock lower down.

Mentor: What would you experience and how would you explain it?

Maddy: The longer distance the rock falls, the harder it hits my bottom hand!

Mentor: Did you have to throw or push the rock to get it to go down?

Maddy: Of course not! Gravity pulls it, as I said before.

Mentor: So the pull between the rock and the earth moves the rock faster and faster as the rock falls. If we could measure the exact position for the rock starting in our hand, what would describe its relative position after a time of t seconds?

Maddy: $y = -t$.

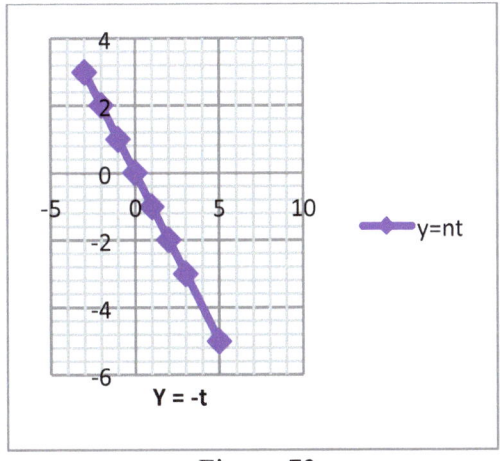

Figure 73

Mentor: So what would the formula for the speed be?

Maddy: The derivative: $y' = -1$.

Mentor: So the speed of the rock is constant -1 ft/sec going down?

Maddy: No! We agreed the rock went faster as it fell.

Mentor: What kind of monomial curve has its tangent, or speed, getting steeper all along its path?

Maddy: The path described by $y = -t^2$ would have the speed formula $y' = -2t$, so the bigger the time the faster the fall!

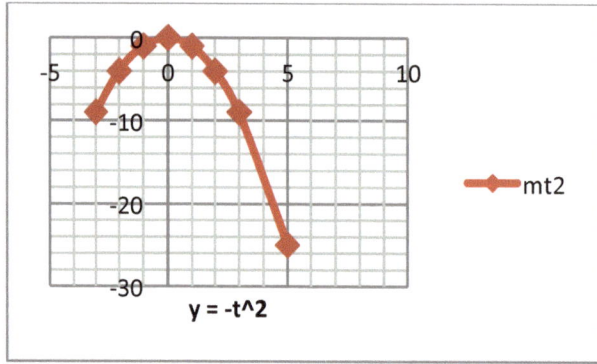
y = -t^2

Figure 74

Mentor: So how fast is your rock falling after 1, 2, or 3 seconds?

Maddy: (1 sec, -2 ft/sec), (2, -4), (3, -6).

Mentor: What can we say about the change in speed? y'', the derivative of the derivative?

Maddy: $y'' = -2$ indicates that there is a constant downward change in the speed, say 2 ft/sec for each second travelled.

Mentor: Looking at the rock as you let it go, how fast is it traveling at the start and at the end of the first second?

Maddy: At the start, when I have just let go, it is not moving. So the speed is zero feet per second. At the end of the first second, we have agreed that the rock will be traveling at 2 feet per second.

Mentor: So how far will the rock fall during the first second after you let go of it?

Maddy: I want to say 2 feet, but I know it was not moving that fast at the beginning. The rock was not moving as fast as 2 feet per second for the whole second, so it must have been traveling very slowly at the start and speeded up as it moved along.

Mentor: Now we have a mystery: barely moving at the start, only moving as fast as 2 feet per second for an instant at the end of the second. How are we to find out how far the rock moved during the entire second?

Maddy: After one tenth of this first second, the rock will be moving at one-tenth of its final speed, $0.1 \times (-2) = -0.2$ feet per second.

Mentor: Looking at the other end of this first second, how fast does your formulation say the rock is moving one-tenth of the second before the end of the second?

Maddy: I suppose the rock would be going nine-tenth of its final speed:
$$0.9 \times (-2) = -1.8 \text{ feet per second.}$$

Mentor: How could we combine these two observations?

Maddy: Maybe we should average these two speeds to get the two speeds balanced out, giving us $\frac{-o.2 + (-1.8)}{2} = \frac{-2}{2}$ or -1 foot per second. Now we can conclude that the rock fell, not 2 feet during its first second of travel but 1 foot from its starting position.

Mentor: Now you have told me that the position of the rock after each second might be given by the equation $y = -t^2$. This means that the speed of the rock, $y' = -2t$, would increase by twice the number of seconds, thus giving us an acceleration, $y'' = -2$ feet per second for every second. Does this correspond to the distance you got by averaging early and late speeds during the first second?

Maddy: No! I have to change my formula to compensate for the changing speed during any time period by multiplying by one half:

$$y = -\frac{1}{2} t^2$$

Mentor: So tell us what happens if you throw the rock upwards at a speed of 6 ft/sec. Does it just keep going?

Maddy: Of course, not! It slows down, stops and falls back to my hand, or the ground.

Mentor: So if gravity pulls the rock toward the ground, $y = -0.5 \, t^2$, and the rock goes up initially, say $y = 6t$, how would you create one equation expressing this idea?

Maddy: I would add them, making

$$y = 6t - \frac{1}{2}t^2$$

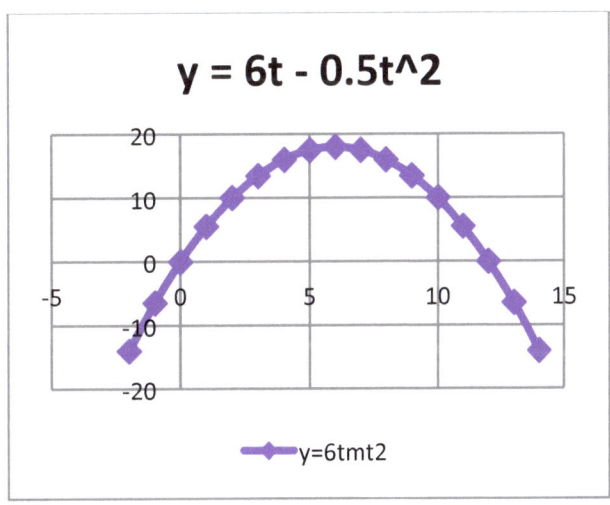

Figure 75

Mentor: When would your rock stop going up and start coming down?

Maddy: In 6 seconds!

Mentor: How did you figure this?

Maddy: Looking at the graph and knowing the parabola is symmetric and the upward motion stops at the halfway point between 0 and 12. So 6 seconds!

   Alternatively, I could take the derivative, $y' = 6 - t$, and set the $y' = 0$, where there is no change in height. This gives me:

$$6 - t = 0,$$
$$\text{or}$$
$$t = 6 \text{ seconds.}$$

Mentor: Now that you have reasoned with your own numbers, we will look at several equations using the English feet or the metric system with meters.

Between the Scenes

I) Using the English system all the measurements are in feet and seconds. Given the equation below, calculate:
a) the speed of an object going straight up at times t = 0, 1, 2, 3, 4, and 5,
b) the maximum height and
c) the acceleration due to gravity.

$$y = \frac{1}{2}(-32)t^2 + 80t + 1$$

II) Using the metric system (with gravity equal 10 meters instead of 9.8 meters per second squared), make the same calculations indicated in part I for the equation given below.

$$y = \frac{1}{2}(-10)t^2 + 70t + 2$$

III) Find the "anti-derivative" of each of the following polynomial derivatives.
Example: If $y' = 4x^3$, then $y = x^4 + 17$ or more generally $y = x^4 + k$, k a constant.
a) $y' = 3x^2$
b) $y' = 0$
c) $y' = 5$
d) $y' = 4x + 1$
e) $y' = 12x^2 + x$
f) $y' = 10x^2 + 7$
g) $y' = 12x^3 + 2x - 8$

Act 8   Area and Integration

Scene 1   Sums of Sequences of Natural Numbers

Mentor: In Scene 6 of Act 10 in <u>You Can Count On It</u>, we saw that there was a formula for adding a sequence of natural numbers from 1 to any number n. Knowing how to find this formula increased our computation speed considerably. Instead of adding $1 + 2 = 3$, and then adding $3 + 3 = 6$, and so on, we learned the formula by stacking the rods in reverse order on top of the steps. Then we could calculate the sum of the area of the stairs as half of the area of the rectangle so formed.

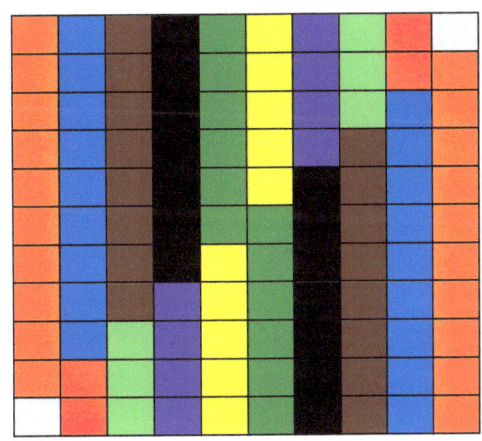

Figure 76

So we have the sum of the first n natural numbers equals $\frac{1}{2} \times L \times W$, where L represents length and W represents width.

In this act we want to calculate the area under a parabola, $y = x^2$, from the origin out to $x = 1$ or $x = 2$. Does anybody have an idea?

Emma: Why don't we do the same thing with parabolas that we did with our stairs? Add the shorter and the longer numbers.

Mentor: OK. Let's see how that works out in calculating the area under $y = x^2$ from 0 to 1. How many divisions are you going to make?

Emma: Let's do what we did with the Cuisenaire Rods, dividing the distance from 0 to 1 into 10 equal parts: 0.1, 0.2, 0.3, 0.4, 0.5, 0.6, 0.7, 0.8, 0.9, 1.0. These are the values: 0.01, 0.04, 0.09, 0.16, 0.25, 0.36, 0.49, 0.64, 0.81, 1.0.

Mentor: How will it be possible to make stairs without using a hundred divisions of each height?

Emma: Let's round every value to the nearest one tenth, giving us:

          0.1, 0.2, 0.3, 0.4, 0.5, 0.6, 0.7, 0.8, 0.9, 1.0
          0.01, 0.04, 0.09, 0.16, 0.25, 0.36, 0.49, 0.64, 0.81, 1.0
          0.00, 0.00, 0.10, 0.20, 0.30, 0.40, 0.50, 0.60, 0.80, 1.0

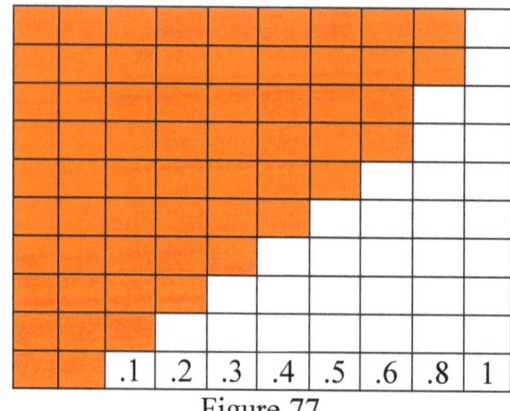

Figure 77

We can see that the white part is not half of the whole square. Maybe turning the white part upside down and around, we can see how much is missing. Even adding back in some of the one-hundredths, we are not going to fill the gap.

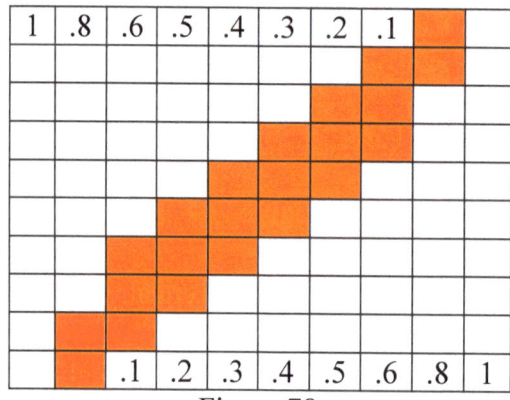

Figure 78

Mentor: Perhaps we could look at this sequence of sums in another form, adding the respective squares to see if the curve formed by the sum is easy to describe.
Emma: We now have $1 + 100 = 101$, $4 + 81 = 85$, $9 + 64 = 73$, $16 + 49 = 65$, $25 + 36 = 61$ and the rest is just the reverse.

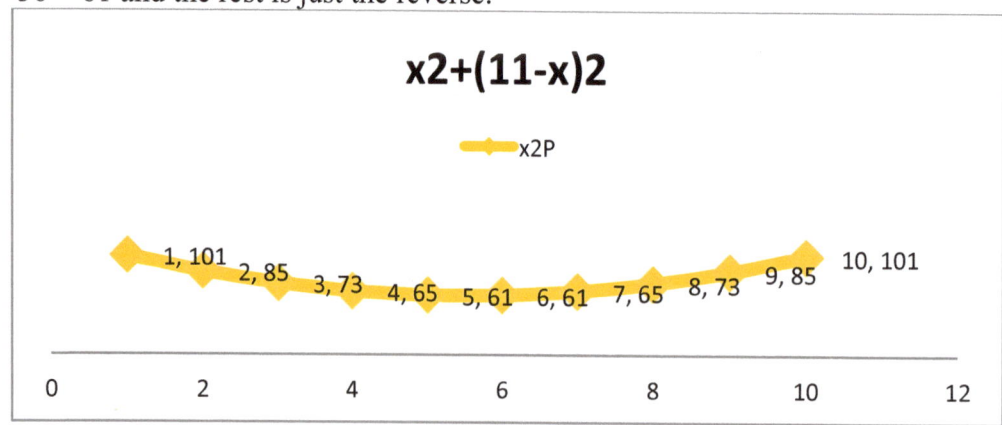

Figure 79

Mentor: Are you prepared to describe this pattern?

Emma: It looks like a piece of a parabola, but I cannot think of a pattern that fits it without a little more thought.

Mentor: What we are looking at is a curve that added to itself upside down, does not give us a simple straight line across the top. The story is a little more complicated. Let me show you a method I learned from Polya's <u>Mathematical discovery</u>, (volume 1, Wiley). In Act 5 Scene 6, we have seen how we can we can write the expansion of a binomial squared, cubed, etc., and now we can use it to see a discovery method. Can anybody write the expansion of $(x + 1)^3$?

Trix: Yes. This is our first use of Pascal's triangle from Act 5 Scene 6.

$$(x + 1)^3 = (x + 1)(x^2 + 2x + 1),$$

multiplying by x first,

$$(x + 1)(x^2 + 2x + 1) = x^3 + 2x^2 + x + x^2 + 2x + 1,$$

collecting like terms we have

$$(x + 1)^3 = x^3 + 3x^2 + 3x + 1$$

Mentor: It is not immediately obvious how we can use this equation. So let me demonstrate what Polya is suggesting. I am going to subtract the x^3 from the right side so all the cubic expressions are on the left.

$$(x + 1)^3 - x^3 = 3x^2 + 3x + 1$$

I am going to let x = 10 instead of 1, so we can avoid worrying about decimal values until we have discovered the right formula. Now we have:

$$(10 + 1)^3 - 10^3 = 3 \times 10^2 + 3 \times 10 + 1,$$
and substituting with 9 instead of 10,
$$(9 + 1)^3 - 9^3 = 3 \times 9^2 + 3 \times 9 + 1, \text{ again with 8}$$
$$(8 + 1)^3 - 8^3 = 3 \times 8^2 + 3 \times 8 + 1$$

Continuing this arrangement until the last x = 1, we obtain:

$$(10 + 1)^3 - 10^3 = 3 \times 10^2 + 3 \times 10 + 1$$
$$(9 + 1)^3 - 9^3 = 3 \times 9^2 + 3 \times 9 + 1$$
$$(8 + 1)^3 - 8^3 = 3 \times 8^2 + 3 \times 8 + 1$$
$$(7 + 1)^3 - 7^3 = 3 \times 7^2 + 3 \times 7 + 1$$
$$(6 + 1)^3 - 6^3 = 3 \times 6^2 + 3 \times 6 + 1$$
$$(5 + 1)^3 - 5^3 = 3 \times 5^2 + 3 \times 5 + 1$$
$$(4 + 1)^3 - 4^3 = 3 \times 4^2 + 3 \times 4 + 1$$
$$(3 + 1)^3 - 3^3 = 3 \times 3^2 + 3 \times 3 + 1$$
$$(2 + 1)^3 - 2^3 = 3 \times 2^2 + 3 \times 2 + 1$$
$$(1 + 1)^3 - 1^3 = 3 \times 1^2 + 3 \times 1 + 1$$

Our final ploy will be to add each of the columns of like expressions. Can anybody see what will happen?

Trix: Yes! I see that 10^3 and $(9+1)^3$ are the same thing, so they will cancel each other. So everything on the left side of the equality will cancel except for the $(10+1)^3 - 1^3$.

Mentor: Excellent observation. Now what happens on the right side of the sum?

Emma: We have 3 times all the squares we want to sum, but do not know the formula, and we have 3 times 10 down to 1, that we do know how to add. And finally we have 10 1's that add up to 10.

Mentor: You are on to the trick or <u>device you use twice</u>. What is the total on the left side?

Emma: $11^3 - 1^3$.

Mentor: And on the right side?

Trix: Working from right to left, we have the sum of 10 1's, or $10 \times 1 = 10$. In the middle we have 3 times the sum of the first 10 natural numbers that we know equals 55. Finally, we have 3 times the squares whose sum we do not know.

Mentor: You have identified all the elements. Let me suggest, following Polya again, that we name the sums again right to left, S_0, S_1, and S_2. So now our equation looks like this:

$$(10+1)^3 - 1^3 = 3 \times S_2 + 3 \times S_1 + S_0$$

What can we do with this expression?

Emma: We can subtract everything from the right side except the sum we want to know.

$$(10+1)^3 - 1^3 - 3 \times S_1 - S_0 = 3 \times S_2$$

This will give us: $(10+1)^3 - 1 - 3 \times 55 - 10 = 3 \times S_2$

Mentor: May I suggest that we replace the 10's with n's, so we can see the pattern evolving? The 55 in the last equation equals $(10 \times 11)/2$.

Trix: Then the equation would be

$$(n+1)^3 - 1^3 - 3 \times \frac{n(n+1)}{2} - n = 3 \times S_2$$

Expanding $(n+1)^3$ we have:

$$n^3 + 3n^2 + 3n + 1 - 1 - \frac{3n(n+1)}{2} - n = 3 \times S_2$$

Clearly the $1 - 1$ goes to zero.

$$n^3 + 3n^2 + 3n - \frac{3n(n+1)}{2} - n = 3 \times S_2$$

Now we use the common denominator of 2 on the left side.

$$\frac{2n^3 + 6n^2 + 6n - 3n^2 - 3n - 2n}{2} = 3 \times S_2$$

106

Simplifying the numerator of the fraction, we get:

$$\frac{2n^3 + 3n^2 + n}{2} = 3S_2$$

Dividing both sides by 3,

$$\frac{2n^3 + 3n^2 + n}{6} = S_2$$

Now we can factor the numerator.

$$\frac{n(2n^2 + 3n + 1)}{6} = S_2$$

And again!

$$\frac{n(n + 1)(2n + 1)}{6} = S_2$$

Mentor: Good persistence! We now have a formula for the sum of n squares from 1 to n. It is a cubic curve, not a quadratic one.

Between the Scenes
1.   Find the formula for the sum of the first n cubed natural numbers.
2.   Find the formula for the sum of the first n natural numbers raised to the fourth power.

Act 8   Scene 2   Hidden Decimal Factor

Mentor: We have seen a variety of factoring problems in arithmetic and elementary algebra. For example, you have seen the area of a garden made up of two plots, each 5 yards on a side but differing in length, one having a length of 8 yards and the other a length of 10 yards. What are the two ways of representing the total area of the garden?
Madeleine: We can add the areas of each smaller garden or we can add the two lengths before multiplying the whole by 5 yards: $5 \times 8 + 5 \times 10 = 5 \times (8 + 10)$.

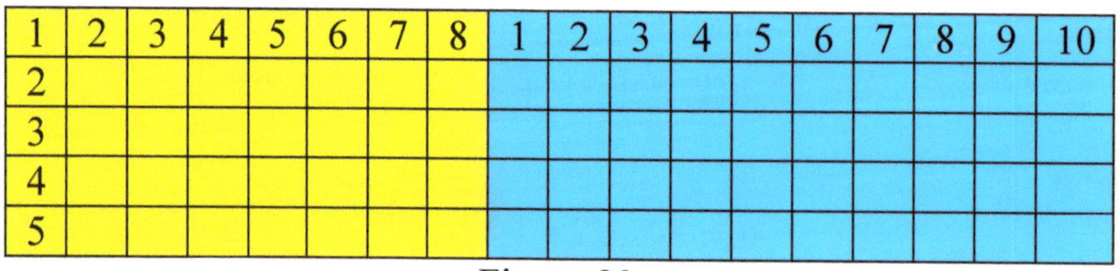

1	2	3	4	5	6	7	8	1	2	3	4	5	6	7	8	9	10
2																	
3																	
4																	
5																	

Figure 80

Mentor: How could we express this, the so-called <u>distributive property</u> of multiplication over addition, using just letters?
Ursula: I would make 'a' the common width and 'b' and 'c' the lengths, giving me the following equation: $a(b + c) = ab + ac$.
Mentor: Toward the end of the course, we will find ourselves interested in calculating areas with decimal additions that we would like to convert to integer additions. A very simple example would be to add: $0.1 + 0.2 + 0.3$. Is there a way we can convert this addition of decimals into a sum of integers, leaving the decimal part for separate calculation?
Maddy: I see this expression as being 0.1 times $1 + 2 + 3$.
$$0.1 \times (1 + 2 + 3)$$
Mentor: That was fast! Now how can we convert an expression like
$$0.1(0.1 + 0.2 + 0.3)$$
into an expression adding just integers: $1 + 2 + 3$?
Trix: I would multiply the expression inside the parentheses by 10 and divide the common factor by 10 to keep the same ultimate value!
$$0.1(0.1 + 0.2 + 0.3) = (0.1)^2(1 + 2 + 3)$$
Mentor: So why call the expression $0.01 + 0.02 + 0.03$ a hidden factoring problem?
Piper: Because multiplication is not obviously shown in the expression.
Mentor: So now how can we convert a decimal expression such as
$$0.1 + 0.2 + 0.3 + \ldots + 10.0$$
into a simpler addition problem, saving the decimal business until the end?
Georgia: $0.1 + 0.2 + 0.3 + \ldots + 10.0 = 0.1 (1 + 2 + 3 + \ldots + 100)$. We know that the sum $1 + \ldots + 100 = (100 \times 101)/ 2$. This makes the calculation of the sum more straight-forward.

Between the Scenes
Find the sum: $0.1(0.1^2 + 0.2^2 + 0.3^2 + \ldots + 1.0^2)$, using the sum of a series of integers squared.

108

Act 8   Scene 3   Approximating Areas Under a Quadratic Curve

Mentor: We have seen how we can calculate the sum of n natural numbers from 1 to n squared. But how does this help us solve our question about the area under the $y = x^2$ curve that we posed in Scene 1? How could we approximate the area under $y = x^2$ from 0 to 1?

Georgia: I would make steps like the ones we made with the rods and see what we get with lots of narrow steps.

Mentor: Good idea. How about starting with just two steps on the interval from 0 to 1?

Georgia: OK! Then I would make each of the two steps half a unit wide as in the diagram below.

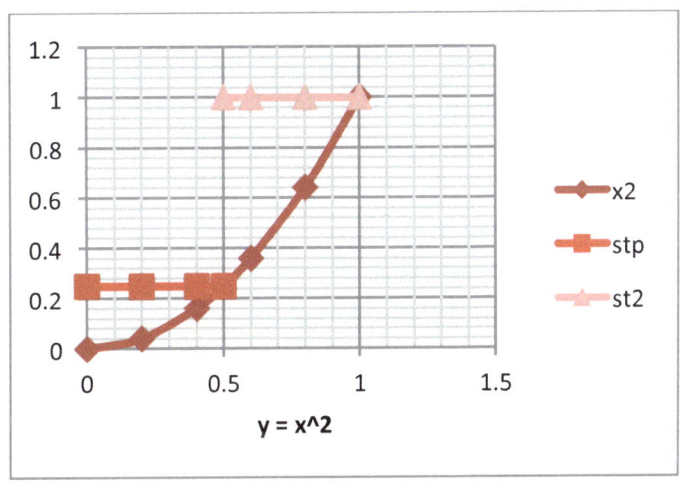

Figure 81

If we picture the two horizontal lines, stp and st2, as the tops of two rectangles, we can calculate their areas simply as width times height: A = 0.5 × 0.25 + 0.5 × 1,

or

$= 0.5 \times 1.25$

$= 0.625$

Of course, I recognize that the calculated area is much more than the area <u>under</u> the $y = x^2$ curve!

Mentor: Can you suggest another way to measure the area under the curve?

Georgia: Yes. Instead of using the height at the right end of each segment, I could use the height at the left end, the lowest value. I imagine moving the red line segment to the right, so it stretches from 0.5 across to 1.0. This area would be B = 0.5 × 0.25 or 0.125.

Mentor: You said the first area was much more than the area under the curve. What can you say about this new area?

Georgia: It is much less than the area we seek.

Mentor: Is there some way we might combine these two calculations to get a better approximation?

Georgia: Well, we could average them. That would give us a number in between, namely (0.625 + 0.125)/2 = 0.750/2 or 0.375

Mentor: How could we improve upon this estimate?

Ursula: We could use more steps. Let's make four steps instead of just 2. See my Figure 82.

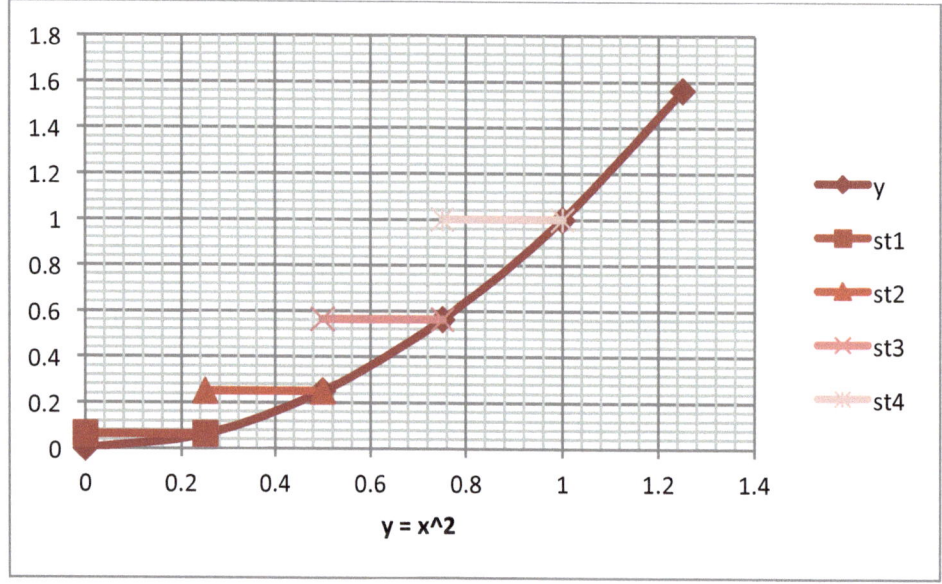

Figure 82

Mentor: What are your calculations now?

Ursula: I'm going to call my sum 'A' for rectangles extending Above the parabola.

$$A = 0.25(0.25)^2 + 0.25(0.5)^2 + 0.25(0.75)^2 + 0.25(1)^2$$

Mentor: Can you see a way to simplify some of the repetitions here?

Ursula: Yes. I can factor out 0.25 from each of the four terms.

$$A = 0.25[(0.25)^2 + (0.5)^2 + (0.75)^2 + (1)^2]$$

Mentor: Do you find another similar repetition, albeit a bit hidden?

Ursula: There is a 0.25-squared common to each of the four terms.

$$A = 0.25[(0.25 \times 1)^2 + (0.25 \times 2)^2 + (0.25 \times 3)^2 + (0.25 \times 4)^2]$$

Factoring out the 0.25-squared, we get

$$A = (0.25)^3(1^2 + 2^2 + 3^2 + 4^2)$$

Mentor: So what does this expression give us to work with?

Ursula: We have the sum of the first four natural numbers squared! I know the formula from our previous work.

$$A = 0.25^3 \times \frac{4 \times (4+1) \times (2 \times 4+1)}{6}$$

Now the factor of 2 in 4 and in 6 reduce, and similarly the factor of 3 in 9 and in 6 goes out, leaving us with a simpler expression.

$$A = 0.25^3 (2 \times 5 \times 3) \text{ or } 0.25^3 (30).$$

So my calculator gives us

$$A = 0.46875$$

Mentor: How to persist! Is this answer better than the one Georgia had us calculate?

Ursula: No, because her final number was the average of the Above approximation and the Below approximation! Now using the formula makes the process much easier. The small approximation we calculate from the left side of four steps starting with the origin for a height of zero.

$$B = (0.25)^3(0^2 + 1^2 + 2^2 + 3^2)$$

By the formula, we now have:

$$B = 0.25^3 \times \frac{3 \times (3+1) \times (2 \times 3+1)}{6},$$

Simplifying to

$$B = 0.25^3 \times 2 \times 7$$

With the 6 in the denominator canceling both the 3 and the 4 in the numerator, leaving just 2, and the 7. Now we have the lower approximation, from my calculator, of:

$$B = 0.21875$$

Now I can average the lower and the upper approximation and get:

$$\text{Average} = 0.34375$$

Mentor: Good job! How do you think we could get an even better approximation?

Piper: I think we should divide the segment into more rectangles, say 10 of them.

Mentor: What would the Above area be?

Piper: Every rectangle would be 0.1 unit wide and $0.1^2, 0.2^2, \ldots, 1.0^2$ units high, but we have learned to factor out the 0.1^2 from each height, giving us the following:

$$A = 0.1 \times [0.1^2 + 0.2^2 + 0.3^2 + \ldots + 1.0^2]$$

or

$$A = 0.1^3 \times [1^2 + 2^2 + 3^2 + \ldots + 10^2]$$

Now we use our formula: The sum of the first n natural numbers squared equals:

$$\text{Sum} = \frac{n(n+1)(2n+1)}{6}, \text{ or } \frac{10 \, (10+1)(2 \times 10+1)}{6}.$$

Simplifying by reducing a common factor of 2 from 10 and 6, and a common factor of 3 from 21 and 6, we are left with

$$\text{Sum} = 5 \times 11 \times 7, \text{ or } 385$$

If we did the same thing with the Below approximation, we would get

$$B = 0.1^3 \times [0^2 + 1^2 + \ldots + 9^2],$$
$$\text{With the Sum} = \frac{9 \times 10 \times 19}{6}$$
$$\text{reduced to}$$
$$3 \times 5 \times 19$$
$$\text{or}$$
$$285$$

If we average the A and B values, we have $(385 + 285) / 2 = 335$ multiplied by the cube of 0.1. So our average value is 0.335.

Mentor: Can you think of a common fraction this average approximates?

Piper: How about 1/3?

Between the Scenes

What are the average values we get if we divide the region into a) 100 equal subsections and b) 1000 equal subsections? Do your results confirm Piper's finding?

Act 8   Scene 4   Area by Anti-derivatives

Mentor: Now we are pretty convinced that the area from 0 to 1 under the y = x² curve is 1/3, and we are also pretty sure that we do not want to put ourselves through such a lengthy approximation calculation every time. As we have seen in Act 7 Scene 6, it is possible to reason backwards from one curve to another having the first curve as its derivative. Let's take several derivatives in the usual way before heading backwards. What is the derivative of y = x?

Piper: $y' = 1$.

Mentor: y = x + 5?

Emma: $y' = 1$.

Mentor: Can you name another curve with the derivative $y' = 1$?

Trix: y = x + k, where k is any constant.

Mentor: Let's move up the exponent list! $y' = x$?

Macey: y = x² + k.

Mentor: What would the derivative look like?

Macey: Oh! Yes. $y' = 2x$, so we will have to multiply a constant times the x term to cancel out the 2. So now we have $y = \frac{1}{2}x^2 + k$.

Mentor: Recognizing that there may be a large class of curves with a certain derivative, we should try to indicate the possibilities, as we see them. Now what curve, or collection of curves, would have $y' = x^2$?

Ursula: That is easy. We must add 1 to the exponent, making it 3 and divide by 3 to leave $y' = x^2$ and not $y' = 3x^2$, so we have $y = \frac{1}{3}x^3$.

Mentor: Now using the anti-derivative to calculate the area under the curve y = x², from 0 to 1, what do I do?

Maddy: You should anti-differentiate $y' = x^2$, to get $y = \frac{1}{3}x^3$ and evaluate it at x = 1, giving us the value $\frac{1}{3}$. This is a lot easier than adding all those squares!

Mentor: So, now tell us the area under the curve y = x³ from 0 to 2.

Georgia: First we have to anti-differentiate and then evaluate. If $y' = x^3$, then $y = \frac{1}{4}x^4$. Now the area under the fourth degree curve starts at the origin and goes up to 2. The area equals $\frac{1}{4} \times 2^4$. This region has the area 4 square units.

Mentor: Who can tell us how to calculate the area under y = x³ from 1 up to 2?

Madeleine: Subtract the area under the curve from 0 to 1 from the area under the curve from 0 to 2. The area equals $\frac{1}{4}(2^4) - \frac{1}{4}(1^4) = 4 - \frac{1}{4}$ or $\frac{16-1}{4} = 3\frac{3}{4}$.

Mentor: So far, in this course, we have avoided heavy notation, but now it makes sense to introduce the definite integral: $\int_1^2 x^3 dx$, where the dx tells us that x is the variable. (An integral might have several variables and constants, but that comes in a later course.) Now you have traveled from Descartes to Newton and seen the two sides of Calculus: differentiation and integration.

Before and After Questions
Pre-High School Calculus Program

Act 1 Given the four points in the coordinate plane A(0, 3), B(5, 0), C(7, 2), and D(-4, -5),
  a) State the slope of each line: AB and CD,
  b) Write an equation for each line: AB and CD,
  c) State the x- and y-intercepts of each line, and
  d) Find the point of intersection of the two lines AB and CD.

Act 2 Find the sum of the first one hundred natural numbers having a remainder of 2 after being divided by 5. Partial list: 2, 7, 12, 17, 22, …, 497.

Act 3 A woman bikes to her car repair shop at 10 mph and returns home by car at a rate twenty miles per hour faster than she went. If the total travel time was 32 minutes, how far away is the repair shop?

Act 4 Sketch graphs of the following curves with their reciprocals and their negatives. Be prepared to describe the various differences and their causes.
  a) $y = 3(x + 2)(x - 5)$, with $1/y$ and $-y$.
  b) $y = 5(x^2 + 16)$, and $1/y$ and $-1/y$.

Act 5 i) Solve the following quadratic equation in two ways—one by the quadratic formula; the other by completing the square.
$$x^2 + 12x + 31 = 0$$
  ii) Complete one more line in Pascal's triangle, page 77, the one starting with x^7.

Act 6 Sketch the three given curves and describe the differences in them and their reciprocals: i) $y = (x - 5)^3$ ii) $y = (x - 5)^2(x - 7)$, and iii) $y = (x - 5)(x - 7)^2$

Act 7 Graph the secant to the curve $y = x^2$ through the points T(3, 9) and S(4, 16).
1) Calculate the slope of line TS, and the slope of TR, R(3.1, 9.61)
2) Calculate the value of the slope of the tangent line through T(3, 9) using the derivative formula.
3) Using the English system all the measurements are in feet and seconds. Given the equation below, calculate:
  a) The speed of an object going straight up at times t = 0, 1, 2, 3, 4, and 5,
  b) The maximum height and the time the rocket reaches this height.
$$y = \frac{1}{2}(-32)t^2 + 384t + 4$$

Act 8 Calculate the area of the region between x = 2 and x = 3 under the parabola $y = 5x^2$
a) By approximation, and
b) By anti-differentiation.

Bibliography

Burger, Edward B., An Introduction to Number Theory
        The Great Courses, Chantilly, VA 2008

Grim, Patrick, The Philosopher's Toolkit: How to Be the Most Rational Person in Any Room
        The Great Courses, Chantilly, VA 2013

Milne, William J. and Walter F. Downey, New Second Course in Algebra
        American Book Company, NY, 1945

Polya, George, Mathematical Discovery: On understanding, learning, and teaching problem solving, vol. I
        John Wiley & Sons, Inc. 1962

Ripley, Amanda, The Smartest Students in the World and how they got that way
        Simon & Schuster, New York, 2013

Sahlberg, Pasi, Finnish Lessons: What can the world learn from educational change in Finland?
        Teachers College, Columbia University, New York, London, 2011

Sherwood and Taylor, Calculus,
        Third Edition, Prentice Hall, Inc., New York April 1954

Snowdon, David, Aging with Grace
        Bantam Books, New York, May 2002

Uspensky, J. V. and Heaslet, M. A., Elementary Number Theory
        McGraw-Hill, New York and London 1939

Zeitz, Paul, The Art and Craft of Mathematical Problem Solving
        Great Courses, Chantilly, VA 2010

Notation

y' (y prime) refers to the derivative of the curve, or function, y.

$\sum_{i=1}^{i=n} i$, (sigma, Greek) for sum of any expression, here just i, starting with i = 1 and ending with i = n.

$\int_a^b y\,dx$ refers to the definite integral of a curve y = curve, or function, in x from a to b.

www.ingramcontent.com/pod-product-compliance
Lightning Source LLC
Chambersburg PA
CBHW050721180526
45159CB00003B/1095